AF616165

Selected Titles in This Series

26 **Rolf Berndt,** An introduction to symplectic geometry, 2001

25 **Thomas Friedrich,** Dirac operators in Riemannian geometry, 2000

24 **Helmut Koch,** Number theory: Algebraic numbers and functions, 2000

23 **Alberto Candel and Lawrence Conlon,** Foliations I, 2000

22 **Günter R. Krause and Thomas H. Lenagan,** Growth of algebras and Gelfand-Kirillov dimension, 2000

21 **John B. Conway,** A course in operator theory, 2000

20 **Robert E. Gompf and András I. Stipsicz,** 4-manifolds and Kirby calculus, 1999

19 **Lawrence C. Evans,** Partial differential equations, 1998

18 **Winfried Just and Martin Weese,** Discovering modern set theory. II: Set-theoretic tools for every mathematician, 1997

17 **Henryk Iwaniec,** Topics in classical automorphic forms, 1997

16 **Richard V. Kadison and John R. Ringrose,** Fundamentals of the theory of operator algebras. Volume II: Advanced theory, 1997

15 **Richard V. Kadison and John R. Ringrose,** Fundamentals of the theory of operator algebras. Volume I: Elementary theory, 1997

14 **Elliott H. Lieb and Michael Loss,** Analysis, 1997

13 **Paul C. Shields,** The ergodic theory of discrete sample paths, 1996

12 **N. V. Krylov,** Lectures on elliptic and parabolic equations in Hölder spaces, 1996

11 **Jacques Dixmier,** Enveloping algebras, 1996 Printing

10 **Barry Simon,** Representations of finite and compact groups, 1996

9 **Dino Lorenzini,** An invitation to arithmetic geometry, 1996

8 **Winfried Just and Martin Weese,** Discovering modern set theory. I: The basics, 1996

7 **Gerald J. Janusz,** Algebraic number fields, second edition, 1996

6 **Jens Carsten Jantzen,** Lectures on quantum groups, 1996

5 **Rick Miranda,** Algebraic curves and Riemann surfaces, 1995

4 **Russell A. Gordon,** The integrals of Lebesgue, Denjoy, Perron, and Henstock, 1994

3 **William W. Adams and Philippe Loustaunau,** An introduction to Gröbner bases, 1994

2 **Jack Graver, Brigitte Servatius, and Herman Servatius,** Combinatorial rigidity, 1993

1 **Ethan Akin,** The general topology of dynamical systems, 1993

An Introduction to Symplectic Geometry

An Introduction to Symplectic Geometry

Rolf Berndt

Graduate Studies in Mathematics

Volume 26

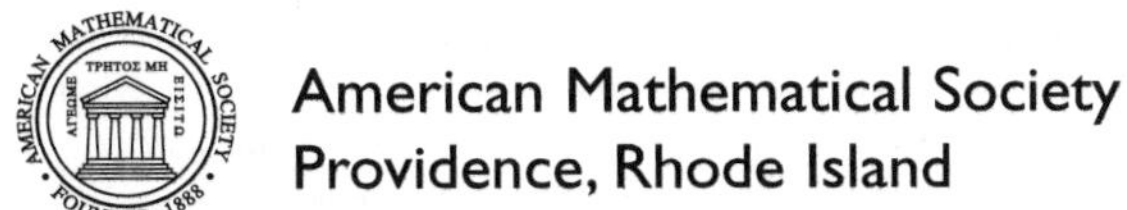

American Mathematical Society
Providence, Rhode Island

2000 *Mathematics Subject Classification.* Primary 53C15, 53Dxx, 20G20, 81S10.

Originally published in the German language by Friedr. Vieweg & Sohn Verlagsgesellschaft mbH, D-65189 Wiesbaden, Germany, as "Rolf Berndt: Einführung in die Symplektische Geometrie. 1. Auflage (1st edition)" © by Friedr. Vieweg & Sohn Verlagsgesellschaft mbH, Braunschweig/Wiesbaden, 1998.

Translated from the German by Michael Klucznik

Abstract. The notions of symplectic form, symplectic manifold and symplectic group appear in many different contexts in analysis, geometry, function theory and dynamical systems. This book assembles tools from different mathematical regions necessary to define these notions and to introduce their application. Among the topics treated here are

- symplectic and Kähler vector spaces,
- the symplectic group and Siegel's half space,
- symplectic and contact manifolds, the theorem of Darboux,
- methods of constructing symplectic manifolds: Kähler manifolds, coadjoint orbits and symplectic reduction,
- Hamiltonian systems,
- the moment map,
- and a glimpse into geometric quantization (in particular the theorem of Groenewold and van Hove) leading to some rudiments of the representation theory of the Heisenberg and the Jacobi group.

The goal of the book is to provide an entrance into a fascinating area linking several mathematical disciplines and parts of theoretical physics.

Library of Congress Cataloging-in-Publication Data

Berndt, Rolf.
[Einführung in die symplektische Geometrie. English]
An introduction to sympletic geometry / Rolf Berndt ; translated by Michael Klucznik.
p. cm. — (Graduate studies in mathematics, ISSN 1065-7339 ; v. 26)
Includes bibliographical references and index.
ISBN 0-8218-2056-7 (alk. paper)
1. Symplectic manifolds. 2. Geometry, Differential I. Title. II. Series.
QA649.B47 2000
516.3′6—dc21 00-033139

∞ The paper used in this book is acid-free and falls within the guidelines
established to ensure permanence and durability.
Visit the AMS home page at URL: http://www.ams.org/

10 9 8 7 6 5 4 3 2 14 13 12 11 10 09

Contents

Preface

> Le caractère propre des méthodes de l'Analyse et de la Géométrie modernes consiste dans l'emploi d'un petit nombre de principes généraux, indépendants de la situation respective des différentes parties ou des valeurs relatives des différents symboles; et les conséquences sont d'autant plus étendues que les principes eux–mêmes ont plus de généralité.
>
> from G. DARBOUX: *Principes de Géométrie Analytique*

This text is written for the graduate student who has previous training in analysis and linear algebra, as for instance S. Lang's *Analysis I* and *Linear Algebra*. It is meant as an introduction to what is today an intensive area of research linking several disciplines of mathematics and physics in the sense of the Greek word *συμπλέκειν* (which means *to interconnect*, or *to interrelate* in English).[1] The difficulty (but also the fascination) of the area is the wide variety of mathematical machinery required. In order to introduce this interrelation, this text includes extensive appendices which include definitions and developments not usually covered in the basic training of students but which lay the groundwork for the specific constructions

[1] I want to thank P. Slodowy for pointing out to me that the name *symplectic group*, which eventually gave rise to the term *symplectic geometry*, was proposed by H. WEYL, [**W**], 1938, in his book, *The Classical Groups* (see footnote on p. 165). The symplectic group was also called the *complex group* or an *Abelian linear group*, this last to honor ABEL, who was the first to study them.

needed in symplectic geometry. Furthermore, more advanced topics will continue to rely heavily on other disciplines, in particular on results from the study of differential equations.

Specifically, the text tries to reach the following two goals:

- To present the idea of the formalism of symplectic forms, to introduce the symplectic group, and especially to describe the symplectic manifolds. This will be accompanied by the presentation of many examples of how they come to arise; in particular the quotient manifolds of group actions will be described,

and

- To demonstrate the connections and interworking between mathematical objects and the formalism of theoretical mechanics; in particular, the Hamiltonian formalism, and that of the quantum formalism, namely the process of *quantization*.

The pursuit of these goals proceeds according to the following plan. We begin in Chapter 0 with a brief introduction of a few topics from theoretical mechanics needed later in the text. The material of this chapter will already be familiar to physics students; however, for the majority of mathematics students, who have not learned the connections of their subject to physics, this material will perhaps be new.

We are constrained, in the first chapter, to consider *symplectic* (and a little later *Kähler*) *vector spaces*. This is followed by the introduction of the associated notion of a symplectic group $Sp(V)$ along with its generation. We continue with the introduction of several specific and theoretically important subspaces, the isotropic, coisotropic and Lagrangian subspaces, as well as the hyperbolic planes and spaces and the radical of a symplectic space.

Our first result will be to show that the symplectic subspaces of a given dimension and rank are fixed up to symplectic isomorphism. A consequence is then that the Lagrangian subspaces form a homogeneous space $\mathcal{L}(V)$ for the action of the group $Sp(V)$. The greatest effort will be devoted to the description of the spaces of positive complex structures compatible with the given symplectic structure. The second major result will be that this space is a homogeneous space, and is, for dim $V = 2n$, isomorphic to the Siegel half space $\mathfrak{H}_n = Sp_n(\mathbb{R})/U(n)$.

The second chapter is dedicated to the central object of the book, namely *symplectic manifolds*. Here the consideration of differential forms is unavoidable. In Appendix A their calculus will be given. The first result of this chapter is then the derivation of a theorem by Darboux that says that the symplectic manifolds are all locally equivalent. This is in sharp contrast to the situation with *Riemannian manifolds*, whose definition is otherwise somewhat parallel to that of the symplectic manifolds. The chapter will then take a glance at new research by considering the assignment of invariants to symplectic manifolds; in particular, the symplectic capacities and the pseudoholomorphic curves will be given.

In the course of the second chapter, we will present several examples of symplectic manifolds:

- First, the example which forms the origin of the theory and remains the primary application to physics is the *cotangent bundle* T^*Q of a given manifold Q.
- Second, the general *Kähler manifold.*
- Third, the *coadjoint orbits.* This description of symplectic manifolds with the operation of a Lie group G can be taken as the second major result of this chapter. We describe a theorem of Kostant and Souriau that says that for a given Lie group G with Lie algebra $\mathfrak{g}$ satisfying the condition that the first two cohomology groups vanish, that is $H^1(\mathfrak{g}) = H^2(\mathfrak{g}) = 0$, there is, up to covering, a one–to–one correspondence between the symplectic manifolds with transitive G–action and the G–orbits in the dual space $\mathfrak{g}^*$ of $\mathfrak{g}$. Here we will need several facts from the theory of Lie algebras and systems of differential equations, and we will at least cover some of the rudiments we require. This will then offer yet another means for introducing one of the central concepts of the field, namely the *moment map.* This will, however, be somewhat postponed so that
- In the fourth and last example, *complex projective space* can be presented as a symplectic manifold; this will be seen as a specific example of the third example, as well as the second; that is, as a coadjoint orbit as well as as a Kähler manifold.

As preparation for the higher level construction of symplectic manifolds, Chapter 3 will introduce the standard concepts of a *Hamiltonian vector field* and a *Poisson bracket.* With the aid of these ideas, we can give the Hamiltonian formulation of classical mechanics and establish the following fundamental short exact sequence:

$$0 \to \mathbb{R} \to \mathcal{F}(M) \to \text{Ham } M \to 0,$$

where $\mathcal{F}(M)$ is the space of smooth functions f defined on the symplectic manifold and given the structure of Lie algebra via the Poisson bracket, and Ham M is the Lie algebra of Hamiltonian vector fields on the manifold.

The third chapter continues with a brief introduction to *contact manifolds.* A theory for these manifolds in odd dimension can be developed which corresponds precisely to that of the symplectic manifolds. On the other hand, both may be viewed as pre-symplectic manifolds. Here the connection will be given through the example of a contact manifold as the surface of constant energy of a Hamiltonian system.

The fourth and fifth chapters will be a mix of further mathematical constructions and their physical interpretations. This will begin with the description of the *moment map* attached to the situation of a Lie group G acting symplectically on a symplectic manifold such that every Hamiltonian vector field is global Hamiltonian. This is a certain function

$$\Phi : M \to \mathfrak{g}^*, \ \mathfrak{g} = \text{Lie } G.$$

The most important examples of the moment maps are the Ad^*–equivariant ones, that is, those that satisfy a compatibility condition with respect to the coadjoint representation Ad^*. The first result of Chapter 4 is that for a symplectic form $\omega = -d\vartheta$ and a G invariant 1-form ϑ such an Ad^*–equivariant map can be constructed. This will then be applied to the cotangent bundle T^*Q, as well as to the tangent bundle TQ, where it will turn out that for a regular Lagrangian function $L \in \mathcal{F}(Q)$ the associated moment map is an integral for the Lagrangian equation associated to L. As examples, we will discuss the *linear* and *angular momenta* in the context of the formalism of the moment map, and so make clear the reason for this choice of terminology.

Next, we describe *symplectic reduction.* Here, we are given a symplectic G-operation on M and an Ad^*–equivariant moment map Φ; under some relatively easy–to–check conditions, for $\mu \in \mathfrak{g}^*$, the quotient

$$M_\mu = \Phi^{-1}(\mu)/G_\mu$$

is again a symplectic manifold. This central result of Chapter 4 has many applications, including the construction of further examples of symplectic manifolds (in particular, we obtain other proofs that the projective space $\mathbb{P}^n(\mathbb{C})$ as well as the coadjoint orbits are symplectic). Another application is the result of classical mechanics on the reduction of the number of variables by the application of symmetry, leading to the appearance of some integrals of the motion.

In the fifth and last chapter, we consider *quantization*; that is, the transition from classical mechanics to quantum mechanics, which leads to many interesting mathematical questions. The first case to be considered is the simplest: $M = \mathbb{R}^{2n} = T^*\mathbb{R}^n$. In this case the important tools are the groups $SL_2(\mathbb{R})$, $Sp_{2n}(\mathbb{R})$, the Heisenberg group $\text{Heis}_{2n}(\mathbb{R})$, the Jacobi group $G^J_{2n}(\mathbb{R})$ (as a semidirect product of the Heisenberg and symplectic groups) and their associated Lie algebras. It will follow that quantization assigns to the polynomials of degree less than or equal to 2 in the variables p and q of $\mathbb{R}^{2n}$ an operator on $L^2(\mathbb{R})$ with the help of the Schrödinger representation of the Heisenberg group and the Weil representation of the symplectic group (more precisely, its metaplectic covering). The theorem of Groenewold and van Hove then says that this quantization is *maximal*; that is, it cannot be extended to polynomials of higher degree.

The remainder of the fifth chapter consists in laying the groundwork for the general situation, which essentially follows KIRILLOV [**Ki**]. Here a subalgebra $\mathfrak{p}$, the *primary quantities*, comes into play, which for the case of $M = T^*Q$ turns out to be the arbitrary functions in q and the linear functions in p. Here yet more functional analysis and topology are required in order to demonstrate the result of Kirillov that for a symplectic manifold, with an algebra $\mathfrak{p}$ in $\mathcal{F}(M)$ of primary quantities relative to the Poisson bracket, a quantization is possible. That is, there is a map which assigns to each $f \in \mathfrak{p}$ a self–adjoint operator $\tilde{f}$ on Hilbert space $\mathcal{H}$ satisfying the conditions

(1) the function 1 corresponds to the identity $\mathrm{id}_{\mathcal{H}}$,

(2) the Poisson bracket of the two functions corresponds to the Lie brackct of operators, and

(3) the algebra of operators operates irreducibly.

There is a one–to–one correspondence between the set of equivalence classes of such representations of $\mathfrak{p}$ and the cohomology group $H^1(M, \mathbb{C}^*)$.

In the first two appendices, manifolds, vector bundles, Lie groups and algebras, vector fields, tensors, differential forms and their basic handling are covered. In particular, the various derivation processes are covered so that one may follow the proofs in the cited literature. A quick reading of this synopsis is perhaps recommended as an entrance to the second chapter. In Chapter 2 some material about cohomology groups will also be required. The third appendix presents some of the rudiments of cohomology theory. In the final appendix, the central concept of coadjoint orbits is prepared by a consideration of the fundamental concepts and constructions of representation theory.

As already mentioned, somewhat more from the theory of differential equations than is usually presented in a beginner's course on the topic, in particular Frobenius' theorem, is required to fully follow the treatment of symplectic geometry given here. Since in these cases the difficulty is not in grasping the statements, this material is left out of the appendices and simply used in the text as needed, though again without proof.

It is not the intention of this text to compete with the treatment of the classical and current literature over the research in the various subtopics of symplectic geometry as can be found, for example, in the books by

ABRAHAM–MARSDEN [**AM**], AEBISCHER *et al.* [**Ae**], GUILLEMIN–STERNBERG [**GS**], HOFER–ZEHNDER [**HZ**], MCDUFF–SALAMON [**MS**], SIEGEL [**Si1**], SOURIAU [**So**], VAISMAN [**V**], WALLACH [**W**] and WOODHOUSE [**Wo**]. Instead we have tried to introduce the reader to the material in these sources and, moreover, to follow the work contained in, for example, GROMOV [**Gr**] and KIRILLOV [**Ki**]. In the hope that this will provide each reader with a starting point into this fascinating area a few parts of chapters 1, 2, and 4 may be skipped by those whose interests lie in physics, and one may begin directly with the sections on Hamiltonian vectorfields, moment maps and quantization.

This text is, with minor changes, a translation of the book "Einführung in die Symplektische Geometrie" (Vieweg, 1998). The production of this text has only been possible through the help of many. U. Schmickler–Hirzebruch and G. Fischer, on the staff of Vieweg–Verlag, have made many valuable suggestions, as has E. Dunne from the American Mathematical Society. My colleagues J. Michaliček, O. Riemenschneider and P. Slodowy, from the *Mathematische Seminar* of the Universität Hamburg, were always, as ever, willing to discuss these topics. A. Günther prepared one draft of this text, and I. Köwing did a newer draft and also showed great patience for my eternal desire to have something or other changed. I also had very successful technical consultation with F. Berndt, D. Nitschke and R. Schmidt. The last of these went through the German text with great attention and smoothed out at least some of what was rough in the text. I would also thank T. Wurzbacher, W. Foerg-Rob and P. Wagner for carefully reading (parts of) the German text and finding some misprints, wrong signs and other mistakes. The translation was done by M. Klucznik, who had an enormous task in producing very fluent English (at least in my opinion) and a fine layout of my often rather involved German style. It is a great joy for me to thank each of these.

R. Berndt

Chapter 0

Some Aspects of Theoretical Mechanics

Symplectic structures arise in a natural way in theoretical mechanics, in particular during the process of quantization, that is, in the passage from classical to quantum mechanics. In order to motivate the study of symplectic geometry, we will begin with a rough sketch of the relevant physics, although we will not cover all the concepts of this field nor give all of the relevant definitions. As references, one may consult the first chapter of VAISMAN [**V**]. A more complete description of the principles of classical mechanics can be found in WOODHOUSE [**Wo**], in the third chapter of ARNOLD [**A**] and in the third and fifth chapters of ABRAHAM–MARSDEN [**AM**]. A further highly recommendable classical source is the first chapter of SIEGEL–MOSER [**SM**]. For the process of quantization, we refer the reader to §15.4 of KIRILLOV [**Ki**]. It is the goal of this text to later return and cover the topics of this chapter in greater detail.

0.1. The Lagrange equations

The purpose of theoretical mechanics is the discovery of principles which allow one to describe the time development of the state of a physical system. In classical mechanics such a state is given as a point P on an n–dimensional real manifold Q (see Section A.1). Q is called the *configuration space*, and P is described by the local coordinates $q_1, \dots, q_n$, called *position variables*. The time development of the system is then described by the curve $\gamma = P(t)$,

$$t \longmapsto P(t) \quad \text{with} \quad P(t_0) = P^0,$$

or in the local coordinates as

$$t \longmapsto q_i(t) \qquad \text{with} \quad q_i(t_0) = q_i^0, \qquad i = 1, \dots, n.$$

Here physical principles must be found which allow one to give the curve as a solution to a differential equation. The starting point for this determination is the classical mechanical *principle of least action*. For this it is assumed that the system has a *Lagrange function* L of the form

$$L = L(q, \dot{q}, t),$$

which is gotten as the difference of the kinetic and the potential energies,

$$L = E_{\text{kin}} - E_{\text{pot}},$$

which is also written as

$$L = T - V.$$

The principle of least action now says that the change in the system proceeds so that the curve γ minimizes the *path integral*

$$\int_{t_0}^{t_1} L\, dt.$$

The variational calculus now says (see COURANT–HILBERT [**CH**], p. 170) that for the minimum curve $\gamma = q(t)$ the system satisfies the Euler–Lagrange equations

$$\frac{d}{dt}\frac{\partial L}{\partial \dot{q}_i} - \frac{\partial L}{\partial q_i} = 0, \qquad i = 1, \dots, n. \tag{1}$$

This can be seen as a system of ordinary differential equations in a $2n$–dimensional space TQ with local coordinates

$$q_1, \dots, q_n, \dot{q}_1, \dots, \dot{q}_n$$

(which can be understood as the *tangent bundle* over the configuration space Q (see Section A.3)). The desired curve γ on Q is the projection of the solution curve $\widetilde{\gamma}$ of (1) onto TQ.

0.2. Hamilton's equations

Classical mechanics now takes the following formulation: for a given Lagrange function L the coordinates *position* and *velocity*, $(q, \dot{q})$, are replaced by the coordinates *position* and *momentum* (q, p) made possible by the transformation

$$p_i = \frac{\partial L}{\partial \dot{q}_i}, \quad i = 1, \dots, n\,.$$

The basis of this concept is the *Legendre transformation* (see ARNOLD [**A**], p. 61 f.) between tangent and cotangent bundles (see Section A.3)

$$\begin{aligned} TQ &\longrightarrow T^*Q, \\ (q, \dot{q}) &\longmapsto (q, p). \end{aligned}$$

Then the time development described on TQ by the Lagrange function $L = L(q, \dot{q}, t)$ (which we can and will assume to be convex in the second argument: see, for example, ARNOLD ([**A**], p. 65)) is replaced by the *Hamiltonian function* H on *phase space* T^*Q defined by

$$H(p,\, q,\, t) := p\dot{q} - L(q, \dot{q}, t) \qquad \text{with} \qquad p = \frac{\partial L}{\partial \dot{q}},$$

where we have used the usual abbreviated symbols for the n–tuple

$$p = (p_1, \ldots, p_n), \quad \frac{\partial L}{\partial q} = \Big(\frac{\partial L}{\partial q_1}, \ldots, \frac{\partial L}{\partial q_n}\Big), \quad \text{etc.}$$

The Lagrange equations (1) are here translated into *Hamilton's equations*

$$\dot{q} = \frac{\partial H}{\partial p}, \quad \dot{p} = -\frac{\partial H}{\partial q}. \tag{2}$$

Because the total differential of $H = H(p,\, q,\, t)$ (see Section A.4) gives

$$dH = \frac{\partial H}{\partial p} dp + \frac{\partial H}{\partial q} dq + \frac{\partial H}{\partial t} dt$$

and by the definition $H = p\dot{q} - L(q, \dot{q}, t)$, we also get

$$dH = \dot{q} dp - \frac{\partial L}{\partial q} dq - \frac{\partial L}{\partial t} dt.$$

Comparing (1) and $p = \dfrac{\partial L}{\partial \dot{q}}$, we get

$$\dot{q} = \frac{\partial H}{\partial p}, \quad \frac{\partial H}{\partial q} = -\frac{\partial L}{\partial q} = -\dot{p}, \quad \frac{\partial H}{\partial t} = -\frac{\partial L}{\partial t}.$$

Hamilton's equations (2) are now (when H is independent of t) a system of ordinary differential equations, which, given a particular set of initial conditions p^0, q^0, gives a unique curve $\widetilde{\gamma}^*$ in phase space T^*Q whose projection γ onto the configuration space Q solves the original problem.

The Hamiltonian function is also written in the form

$$H = H(p,\, q,\, t) = (T) + V,$$

where V is the potential energy of the system and T is the kinetic energy given in terms of the variables q and p.

0.3. The Hamilton-Jacobi equation

Yet another formulation of the problem passes from the solution of a system of ordinary differential equations to the solution of a partial differential equation. The resulting partial differential equation is the *Hamilton–Jacobi equation*

$$H\left(q, \frac{\partial S}{\partial q}, t\right) + \frac{\partial S}{\partial t} = 0 \tag{3}$$

for the *action function* S. Here, giving a solution which is dependent on t, the n variables q, and the n initial parameters a,

$$S = S(q,\, t,\, a),$$

is equivalent to giving a solution $q = q(t)$, $p = p(t)$ of (2). Here we present only the following consideration:

Let $S = S(q,\, t,\, a)$ be a solution of (3) with

$$\det\left(\frac{\partial^2 S}{\partial q_i\, \partial a_k}\right) \neq 0\,.$$

Then the n equations

$$\frac{\partial S}{\partial a_\ell} = b_\ell\,,\ \ \ell = 1, \ldots, n\,,$$

in the q_i are solvable in the $q_i = \varphi_i(t, a, b)$, $i = 1, \ldots, n$. This allows one to write

$$p_\ell := \frac{\partial S}{\partial q_\ell}$$

as a function of t, a, and b:

$$p_\ell = \psi_\ell(t,\, a,\, b).$$

These q_i, p_i satisfy Hamilton's equations (2), since differentiating

$$H\left(q, \frac{\partial S}{\partial q}(q,\, t,\, a), t\right) + \frac{\partial S}{\partial t} = 0 \tag{+}$$

with respect to a_ℓ gives

$$\sum_k \frac{\partial H}{\partial p_k} \frac{\partial^2 S}{\partial a_\ell\, \partial q_k} + \frac{\partial^2 S}{\partial a_\ell\, \partial t} = 0, \quad \ell = 1, \ldots, n\,.$$

And differentiating $\dfrac{\partial S}{\partial a_\ell} = b_\ell$ with respect to t gives

$$\sum_k \frac{\partial^2 S}{\partial q_k\, \partial a_\ell} \dot{q}_k + \frac{\partial^2 S}{\partial t\, \partial a_\ell} = 0\,.$$

Taking the difference of the two equations yields

$$\sum_k \frac{\partial^2 S}{\partial a_k\,\partial a_\ell}\left(\frac{\partial H}{\partial p_k} - \dot{q}_k\right) = 0, \qquad \ell = 1, \dots, n,$$

and since

$$\det\left(\frac{\partial S}{\partial q\,\partial a}\right) \neq 0$$

we arrive at half of Hamilton's equations. Next, differentiating (+) with respect to q_ℓ gives

$$\frac{\partial H}{\partial q_\ell} + \sum_k \frac{\partial H}{\partial p_k}\,\frac{\partial^2 S}{\partial q_k\,\partial q_\ell} + \frac{\partial^2 S}{\partial q_\ell\,\partial t} = 0,$$

and then, differentiating $p_\ell = \dfrac{\partial S}{\partial q_\ell}$ with respect to t, we get

$$\dot{p}_\ell = \sum_k \frac{\partial^2 S}{\partial q_k\,\partial q_\ell}\,\dot{q}_k + \frac{\partial^2 S}{\partial t\,\partial q_\ell}\,.$$

Taking the difference of these two equations yields, considering the satisfaction of the relation $\dfrac{\partial H}{\partial p} = \dot{q}$,

$$\dot{p}_\ell = -\frac{\partial H}{\partial q_\ell}\,.$$

There is yet another way (which at first glance looks different) to derive the Hamilton–Jacobi equation (see ARNOLD [**A**], pp. 253-5). Here the path integral

$$S_{q^0,t^0}(q,\, t) = \int_\gamma L dt$$

is taken along the curve γ from (q^0, t^0) to $(q,\, t)$ that minimizes the integral, and it is shown that

$$dS = pdq - Hdt.$$

Then it is immediately clear that for S the equations

$$\frac{\partial S}{\partial t} = -H(p, q, t) \quad \text{and} \quad \frac{\partial S}{\partial q} = p$$

hold, and therefore also (3).

0.4. A symplectic interpretation

Here we continue from Section 0.2. The Hamiltonian function H defines a *Hamiltonian vector field* X_H on phase space T^*Q. Relative to the usual coordinates (q, p), this is defined by (see Section A.4)

$$X_H := \sum_i \frac{\partial H}{\partial p_i}\frac{\partial}{\partial q_i} - \sum \frac{\partial H}{\partial q_i}\frac{\partial}{\partial p_i}.$$

Given a vector field X, the question immediately arises as to the existence of *integral curves* γ; that is, curves whose tangent vectors $\dot\gamma(t)$ at every point of $\gamma(t)$ are equal to the given vector of the vector field at that point, which, in symbols then, is $\dot\gamma(t) = X_H(\gamma(t))$. For

$$\gamma(t) = (q(t),\, p(t))$$

this condition leads to Hamilton's equations (2)

$$\frac{\partial H}{\partial p} = \dot q, \quad \frac{\partial H}{\partial q} = -\dot p.$$

With the help of a little bit more from the theory of differential forms (see Section A.4), this can be reformulated as follows: there exist a 2–*form*

$$\omega = \sum_i dq_i \wedge dp_i \in \Omega^2 \quad \text{on } T^*Q$$

and an *inner product* i such that, from any vector field X, the 2–form ω gives us a 1–form $i(X)\omega$. Then Hamilton's equations (2) are equivalent to

$$i(X_H)\omega = dH\,. \tag{4}$$

0.5. Hamilton's equations via the Poisson bracket

On any pair of (arbitrarily often) differentiable functions f, g on phase space T^*Q we may take their *Poisson bracket* $\{\,,\,\}$, [1] defined by the equation

$$\{f,\, g\} := \sum_i \frac{\partial f}{\partial q_i}\frac{\partial g}{\partial p_i} - \frac{\partial f}{\partial p_i}\frac{\partial g}{\partial q_i}, \qquad f, g \in \mathcal{F}(T^*Q)\,.$$

This Poisson bracket endows the space of functions $\mathcal{F}(T^*Q)$ with the structure of a Lie algebra (see Section B.1). This will be discussed in more detail later. Here we only note that Hamilton's equations (2) can be written with the help of the Poisson bracket as

$$\dot q = \{q, H\}\,,\ \dot p = \{p,\, H\}\,. \tag{5}$$

[1] Warning: in the literature (for example, in **[Ki]**) one sometimes takes for $\{f,\, g\}$ the negative of what is taken here!

From this it does not take too much work to see that, generally, for the time development of an observable given by f, the above system must satisfy the condition

$$\dot{f} = \{f, H\}. \tag{5'}$$

0.6. Towards quantization

The term *quantization* will indicate the process by which a *corresponding quantum system* is constructed from a given classical system. Thus we must find a transition from the point in phase space T^*Q which describes the state of a classical system to an element v (more accurately, a 1–dimensional subspace $v\,\mathbb{C}$) of a complex Hilbert space $\mathcal{H}$ with the aid of the probability distribution for the state of a quantum mechanical system. This transition should give, for the Hamilton function H and the classical observables f, corresponding self–adjoint operators $\widehat{H}$ and $\widehat{f}$ in $\mathcal{H}$ for the quantum situation. Thus one would expect that the equation (5′)

$$\dot{f} = \{f, H\},$$

giving the time course in T^*Q in the classical case should pass to an analogous equation on the operators

$$\dot{\widehat{f}} = c\,[\widehat{f}, \widehat{H}],$$

where [,] is here the natural *Lie bracket*,

$$[A, B] = AB - BA,$$

where $c = -\frac{ih}{2\pi}$ (c is, up to the factor i, a factor ensuring the symmetry of the operators, a constant from physics, and h is called *Planck's constant*).

We will later examine for which $f \in \mathcal{F}(T^*M)$ one may define a mapping $f \longmapsto \widehat{f}$ whose images are self–adjoint operators and which satisfies

$$1 \longmapsto \widehat{1} = id_{\mathcal{H}}$$

and

$$\widehat{\{f_1, f_2\}} = c\left[\widehat{f_1}, \widehat{f_2}\right].$$

With this process considerable problems will appear, but we will finally see that, at least for the so-called *primary quantities* (that is, f either a polynomial of degree 2 in q and p or a linear function of the $p_1, \ldots, p_n$ and an arbitrary function of the $q_1, \ldots, q_n$) solutions can be found.

Chapter 1

Symplectic Algebra

The soon to be introduced *symplectic manifolds* can be thought of locally as *symplectic vector spaces.* It is therefore necessary to define and study vector spaces with additional structure. This additional structure is given by

a) a scalar product,

b) a symplectic form,

c) a complex structure.

To begin, we rework some old and trusted linear and multilinear algebra. As a final result, we will describe all spaces with a given symplectic structure compatible with a complex structure as a *Siegel half space.* This space is important not only for geometry, but also for function theory, although as to the latter, we will only be able indicate a small part of this importance. The second chapter of VAISMAN: *Symplectic Geometry and Secondary Characteristic Classes* [**V**], as well as the third chapter of ABRAHAM–MARSDEN [**AM**] serve as good guides to this chapter. For additional background, E. ARTIN's *Geometric Algebra* [**Ar**] can be recommended.

1.1. Symplectic vector spaces

Now we let K be an arbitrary field of characteristic 0. Later, we will restrict to $K = \mathbb{R}$. Also let V be a finite–dimensional K vector space (with $\dim V = p$). Then the basic underlying concept of symplectic space is given by the following definition.

DEFINITION 1.1. A *symplectic form*

$$\omega : V \times V \to K$$

is an antisymmetric and nondegenerate bilinear form; that is, it satisfies

$$\omega(v, v) = 0 \quad \text{for all } v \in V,$$

and if

$$\omega(v, w) = 0 \quad \text{for all } v \in V,$$

then $w = 0$. A vector space V is called a *symplectic vector space* if it is equipped with a symplectic form.

Remark 1.2.

(1) Abraham-Marsden [**AM**] consider infinite–dimensional symplectic vector spaces, as well.

(2) In the case of $K = \mathbb{R}$ this definition is parallel to the definition of a *Euclidean vector space*; that is, to a space having a *scalar product*, a symmetric positive definite bilinear form which is usually denoted by s or by $\langle\ ,\ \rangle$.

Let

$$\underline{e} = (e_1, \ldots, e_p)$$

be a basis of V. Then the bilinear form ω on V in terms of $\underline{e}$ can be given in matrix form by

$$\omega \mapsto \omega_{\underline{e}} = (\omega_{ij}) \in M_p(K) \quad \text{with } \omega_{ij} = \omega(e_i, e_j).$$

For $K = \mathbb{R}$ there is a nice classification of the normal forms of symmetric and skew–symmetric bilinear forms.

Proposition 1.3. *Let V be a p–dimensional $\mathbb{R}$ vector space.*

i) *In the case that s is a symmetric bilinear form of rank r, there is a basis $\underline{e}$ of V relative to which*

$$s_{\underline{e}} = \begin{pmatrix} \varepsilon_1 & & & & & \\ & \ddots & & & & \\ & & \varepsilon_r & & & \\ & & & 0 & & \\ & & & & \ddots & \\ & & & & & 0 \end{pmatrix} \quad \textit{with } \varepsilon_i = \pm 1, \ i = 1, \ldots, r.$$

ii) *In the case that ω is an antisymmetric bilinear form of rank r, then $r = 2n$ and there is a basis $\underline{e}$ of V relative to which*

$$\omega_{\underline{e}} = \begin{pmatrix} 0 & E_n & 0 \\ -E_n & 0 & 0 \\ 0 & 0 & 0 \end{pmatrix} \quad \textit{with } E_n \in M_n(\mathbb{R}) \text{ the unit matrix.}$$

Proof. i) By a process analogous to the Gram–Schmidt orthonormalization, it can be assumed that since s is symmetric, it satisfies the polarization identity

$$s(v, w) = (1/4)\big(s(v+w, v+w) - s(v-w, v-w)\big).$$

Thus for $s \neq 0$, there is an $e_1' \in V$ with $s(e_1', e_1') \neq 0$. e_1' can be multiplied by a scalar, giving e_1 with $\varepsilon_1 := s(e_1, e_1) = \pm 1$. Let

$$V_1 := \mathbb{R}\, e_1 \quad \text{and} \quad V_2 := \{v \in V,\ s(e, e_1) = 0\}.$$

It is then clear that $V_1 \cap V_2 = \{0\}$, and so $V_1 + V_2 = V$; then for $v \in V$,

$$v - \varepsilon_1 s(v, e_1)\, e_1 \in V_2.$$

One may now continue by induction. So long as $s \neq 0$ on $V_2 \times V_2$, we may find an $e_2 \neq 0$ in V_2 with $s(e_2, e_2) = \varepsilon_2 = \pm 1$, and so on.

ii) For $\omega \neq 0$, there must be $e_1, e_{n+1} \in V$ with $\omega(e_1, e_{n+1}) \neq 0$. After multiplying e_1 by a scalar, it can be assumed that $\omega(e_1, e_{n+1}) = 1$. Since ω is skew–symmetric, we have

$$\omega(e_1, e_1) = \omega(e_{n+1}, e_{n+1}) = 0,$$

and the matrix for ω' in the plane E_1 spanned by e_1 and e_{n+1} is

$$\begin{pmatrix} 0 & 1 \\ -1 & 0 \end{pmatrix}.$$

Now let V_2 be the ω–orthogonal complement of E_1, so

$$V_2 := \big\{v \in V \,|\, \omega(v, v_1) = 0 \quad \text{for all } v_1 \in E_1\big\}.$$

We have $E_1 \cap V_2 = \{0\}$ and $V = E_1 \oplus V_2$, and for $v \in V$

$$v - \omega(v, e_{n+1})\, e_1 + \omega(v, e_1)\, e_{n+1} \in V_2.$$

Given that $\omega \neq 0$ on V_2, one may repeat the above procedure for V_2 and obtain e_2 as well as e_{n+2} so that $\omega(e_2, e_{n+2}) = 1$. Inductively we get the claimed matrix $\omega_{\underline{e}}$. □

The statement ii) clearly holds as well for more general fields $K \neq \mathbb{R}$.

Let V^* denote the *dual space* of V and $\underline{e}^*$ the *dual basis* to $\underline{e}$ on V^*, which satisfies

$$e_i^*(e_j) = \langle e_j, e_i^* \rangle = \delta_{ij}.$$

One of the basic results of (multi–)linear algebra (see also Section A.4) is that the space $A^q(V, K)$ of skew–symmetric q–linear functions from V^q to K is isomorphic to the qth exterior product $\Lambda^q V^*$ of V^*. $\Lambda^q V^*$ has as K–basis the set

$$e_{i_1}^* \wedge \ldots \wedge e_{i_q}^* \quad \text{with } i_1 < \ldots < i_q.$$

In particular, an antisymmetric bilinear form ω with the matrix $\omega_{\underline{e}} = (\omega_{ij})$ relative to $\underline{e}$ can also be written as

$$\omega = \sum_{i<j} \omega_{ij}\, e_i^* \wedge e_j^*.$$

Written this way, ω works as a function by sending $(v,\, w) \in V \times V$ to

$$\omega\,(v, w) = \sum_{i<j} \omega_{ij}\big(e_i^*(v)e_j^*(w) - e_i^*(w)e_j^*(v)\big).$$

The statement ii) can be reformulated as

COROLLARY 1.4. *By an appropriate choice of basis $\underline{e}$, the antisymmetric bilinear form ω can be written as*

$$\omega = \sum_{i=1}^{n} e_i^* \wedge e_{i+n}^*.$$

Such a representation will be called the *canonical form* of ω, and $\underline{e}$ a *symplectic basis* of V. Then

$$\omega\,(v,\, v') = \sum_{i=1}^{n}(x_i y_i' - x_i' y_i),$$

when for $v \in V$ the components relative to $\underline{e}$ are given by

$$v = \sum_{i=1}^{n} x_i e_i + \sum_{i=1}^{n} y_i e_{i+n} + \sum_{i=1}^{p-2n} z_i e_{2n+i}.$$

Of particular importance for symplectic geometry is naturally the nondegenerate skew–symmetric bilinear form ω; in this case $p = r = 2n$, so that V must have even dimension. A criterion for $\omega \in \Lambda^2 V^*$ to be nondegenerate is that the nth exterior power $\omega^n = \omega \wedge \ldots \wedge \omega$ of ω must be a nonzero multiple of the *volume form*

$$\tau = e_1^* \wedge \ldots \wedge e_p^* \in \Lambda^p V^*.$$

For τ, we have

$$\tau\,(v_1, \ldots, v_p) = \det(a_{ij}) \quad \text{with } v_i = \sum_{j=1}^{p} a_{ij}\, e_j,\ i = 1, \ldots, p.$$

For $\omega = \sum e_i^* \wedge e_{n+i}^*$, we have, in the usual notation, $\omega^n = n!(-1)^{[n/2]}\tau$, where $[x]$ is the function that returns the largest integer $\leq x$ for all $x \in \mathbb{R}$. In general,

$$\tau_\omega = \frac{(-1)^{[n/2]}}{n!}\,\omega^n$$

defines an *orientation* on V (see ABRAHAM-MARSDEN [**AM**], pp. 165-166).

A symplectic form ω makes possible the identification of

$$\begin{aligned} \omega^b : V &\rightarrow V^*, \\ v &\mapsto \omega^b(v), \end{aligned}$$

with

$$\omega^b(v)(v') = \omega(v, v') \quad \text{for } v, v' \in V.$$

We will use the letter i for a general *inner product*

$$\begin{aligned} i : V \times \Lambda^q V^* &\rightarrow \Lambda^{q-1} V^*, \\ (v, \vartheta) &\mapsto i(v)\vartheta, \end{aligned}$$

where $i(v)\vartheta$ is the $(q-1)$–linear function given by

$$i(v)\vartheta(v_1, \ldots, v_{q-1}) = \vartheta(v, v_1, \ldots, v_{q-1}),$$

and so

$$\omega^b(v) = i(v)\omega \in V^*.$$

It follows easily for ω, as in Corollary 1.4, that

$$\begin{aligned} i(e_j)\,\omega &= \omega^b(e_j) &&= e^*_{j+n}, \qquad j = 1, \ldots, n, \\ i(e_{j+n})\,\omega &= \omega^b(e_{j+n}) &&= -e^*_j. \end{aligned}$$

Although it is almost trivial, we recommend

EXERCISE 1.5. Given a bilinear form ω on V with $\dim V = m$, show that the following are equivalent:

a) ω is non–degenerate,

b) ω^b is an isomorphism,

and, in the case that $m = 2n$,

c) $\omega^n = \omega \wedge \ldots \wedge \omega \neq 0$.

Although these statements are left unproved, we will not hesitate to use them later in the text.

In parallel to the situation in Euclidian geometry, we may form the following definition in symplectic geometry.

DEFINITION 1.6. Two vectors v, w from the symplectic vector space (V, ω) are called *ω–orthogonal, skew–orthogonal* or – when there is no doubt about which ω is intended – simply *orthogonal*, whenever

$$\omega(v, w) = 0.$$

This is also indicated by $v \perp w$.

EXAMPLE 1.7. K^{2n} with the symplectic form ω given by $\omega(v, v') = \sum (x_i y_i' - x_i' y_i)$ for

$$\begin{aligned} v &= x_1 e_1 + \ldots + x_n e_n + y_1 e_{n+1} + \ldots && + y_n e_{2n}, \\ v' &= x_1' e_1 + \ldots && + y_n' e_{2n} \end{aligned}$$

with the canonical basis $\underline{e} = (e_1, \ldots, e_{2n})$ is the *standard symplectic space.*

Given Proposition 1.3, every $2n$–dimensional symplectic space can be described this way.

EXAMPLE 1.8. Let W be an n–dimensional K vector space and W^* its dual space. Then $V = W \oplus W^*$ is a symplectic space with

$$\omega : V \times V \to K$$

given by

$$\omega(t_1 + \tau_1, t_2 + \tau_2) = \tau_1(t_2) - \tau_2(t_1) \quad \text{for } t_1, t_2 \in W, \ \tau_1, \tau_2 \in W^*.$$

Remark 1.9. Relative to this last example, note that a symplectic space V has many decompositions $V = W \oplus W^*$. Let $\underline{e}$ be a symplectic basis for V, such that W is the span of $e_1, \ldots, e_n$. Then W^* is isomorphic, via ω^b, to the subspace spanned by $e_{n+1}, \ldots, e_{2n}$; so $V = W \oplus W^*$ with the form defined in the previous example.

EXAMPLE 1.10. There is for every $k \in \mathbb{N}$ a $p = (2n + k)$–dimensional K vector space U with a skew–symmetric bilinear form $\widehat{\omega} : U \times U \to K$ of rank $2n$. Then the *annihilator* of $\widehat{\omega}$,

$$U_0 := \{u \in U;\ \widehat{\omega}(u, w) = 0 \quad \text{for all } w \in U\},$$

has dimension k, and $\widehat{\omega}$ induces a symplectic form ω on $V := U/U_0$. We will say that the symplectic space (V, ω) is given by the reduction of $(U, \widehat{\omega})$.

1.2. Symplectic morphisms and symplectic groups

Just as in the case of a Euclidian vector space, where the scalar product permits one to define *orthogonal* morphisms, we have, in symplectic geometry, a natural definition of morphism:

DEFINITION 1.11. Let (V_i, ω_i) be two symplectic vector spaces and $\phi : V_1 \to V_2$ a linear map. Then we call ϕ *symplectic* whenever

$$(*) \qquad \omega_2\big(\phi(v), \phi(w)\big) = \omega_1(v, w) \quad \text{for all } v, w \in V_1.$$

Remark 1.12. A symplectic morphism is necessarily injective, since if $\phi(v) = 0$, then $(*)$ forces $v = 0$, since ω_1 is non-degenerate. For $\dim V_1 = \dim V_2 < \infty$, ϕ must therefore be an isomorphism. Symplectic isomorphisms will be called *symplectomorphisms.*

When $(V_1, \omega_1) = (V_2, \omega_2) = (V, \omega)$, any symplectic map ϕ must be an automorphism of (V, ω). The collection of symplectic automorphisms forms a group under the usual composition, called the *symplectic group* of (V, ω) and denoted $Sp(V)$. In particular, for $V = K^{2n}$ with the standard form, this group will be written as $Sp_n(K)$ (unfortunately it is also written as $Sp_{2n}(K)$ in the literature!).

The elements $M \in Sp_n(K)$ are naturally matrices from $GL_{2n}(K)$. They can be written in the following way. For the standard form, we have

$$\omega(v, v') = \sum (x_i y_i' - x_i' y_i),$$

which, with the use of the matrix

$$J = J_n = \begin{pmatrix} 0 & E_n \\ -E_n & 0 \end{pmatrix}$$

and column vectors v, v' with ${}^t v = (x_1, \ldots, x_n, y_1, \ldots, y_n)$, can also be written as

$$\omega(v, v') = {}^t v J v'.$$

The matrix M leaves this form invariant, that is,

$$\omega(Mv, Mv') = \omega(v, v'),$$

exactly when

$$(+) \qquad {}^t M J M = J.$$

Then from simple computation, we get

Remark 1.13. For $A, B, C, D \in M_n(K)$, the following are equivalent:

i) $$M = \begin{pmatrix} A & B \\ C & D \end{pmatrix} \in Sp_n(K),$$

ii) $${}^t AC = {}^t CA, \ {}^t BD = {}^t DB, \ {}^t AD - {}^t CB = E_n,$$

iii) $$A^t B = B^t A, \ C^t D = D^t C, \ A^t D - B^t C = E_n.$$

Special symplectic matrices are J as well as

$$U_V = \begin{pmatrix} V & 0 \\ 0 & V^* \end{pmatrix} \qquad \text{with } V^* = ({}^tV)^{-1},$$

and

$$T_S = \begin{pmatrix} E & S \\ 0 & E \end{pmatrix} \qquad \text{with } {}^tS = S.$$

PROPOSITION 1.14. *These matrices generate $Sp_n(K)$.*

Proof. See EICHLER [**E**], p. 47. □

As a consequence of this proposition, we immediately get that

$$\det M = 1 \qquad \text{for } M \in Sp_n(K),$$

since this is clearly the case for the generators. This statement can also be derived from the fact that a symplectic automorphism ϕ relative to ω must also fix ω^n, the volume form. Therefore, we have

$$\begin{aligned} e_1^* \wedge \ldots \wedge e_{2n}^*(\phi e_1, \ldots, \phi e_{2n}) &= (\det \phi)\, e_1^* \wedge \ldots \wedge e_{2n}^*(e_1, \ldots, e_{2n}) \\ &= e_1^* \wedge \ldots \wedge e_{2n}^*(e_1, \ldots, e_{2n}), \end{aligned}$$

and thus $\det \phi = 1$.

PROPOSITION 1.15. *Let $M \in Sp_n(K)$ and λ an eigenvalue of M with multiplicity k. Then $1/\lambda$ is also an eigenvalue with multiplicity k.*

Proof. Consider

$$P(t) = \det(M - tE_{2n}),$$

the characteristic polynomial of M. Then by using (+) and the fact that $\det M = 1$, we have

$$\begin{aligned} P(t) &= \det({}^tM - tE_{2n}) = \det\left(J^{-1}({}^tM - tE_{2n})J\right) \\ &= \det(M^{-1} - tE_{2n}) = \det M^{-1} \det(E_{2n} - tM) \\ &= t^{2n} \det(M - (1/t)E_{2n}). \end{aligned}$$

□

Remark 1.16. For $K = \mathbb{R}$, if $M \in Sp_n(K)$ as a complex matrix has the eigenvalue $\lambda \in \mathbb{C}$, then M also has the eigenvalues $\overline{\lambda}, 1/\lambda$ and $1/\overline{\lambda}$.

These statements about eigenvalues are fundamental to the qualitative theory, including that of the stability of Hamiltonian systems. Here we give just a few comments on the topic of *stability* (see ARNOLD [**A**], p. 227):

DEFINITION 1.17. A morphism ϕ of V is called *stable* if for each $\varepsilon > 0$ there is a $\vartheta > 0$ such that

$$\| \phi^N v \| < \varepsilon, \quad \text{for all } N \in \mathbb{N}, \quad \text{as soon as} \quad \| v \| < \vartheta.$$

EXERCISE 1.18. If $\phi \in Sp(V)$ has an eigenvalue λ with $|\lambda| \neq 1$, then ϕ is not stable.

EXERCISE 1.19. Whenever the eigenvalues λ of $\phi \in \text{End } V$ are distinct and have norm 1, the transformation is stable.

DEFINITION 1.20. $\phi \in Sp(V)$ is called *strongly stable* if every neighbor $\phi_1 \in Sp\,(V)$ (that is, given a fixed basis for V, the matrix entries of ϕ_1 differ from those of ϕ by less than a fixed ε) is stable.

EXERCISE 1.21. If for $\phi \in Sp\,(V)$ all $2n$ eigenvalues are distinct and have norm 1, ϕ is strongly stable.

The symplectic group is a significant object in both function theory and algebraic geometry. For clarification the reader may wish to consult SIEGEL [**Si1, Si2**]. $Sp_n(\mathbb{R})$ is a Lie group and has as associated Lie algebra $\mathfrak{sp}_n(\mathbb{R}) = \text{Lie } Sp_n(\mathbb{R})$ (see Section B.2). It is easy to see that with the standard form ω we have

$$\mathfrak{sp}_n(\mathbb{R}) = \{M \in M_{2n}(\mathbb{R}),\ \omega\,(Mv,\, w) + \omega\,(v,\, Mw) = 0 \quad \text{for all } v,\, w \in \mathbb{R}^{2n}\}$$

$$= \{M \in M_{2n}(\mathbb{R}),\ {}^tMJ + JM = 0\}.$$

Here, as in the last proposition, we have for $M \in \mathfrak{sp}_n(\mathbb{R})$ that if λ is an eigenvalue then so is $-\lambda$, and both have the same multiplicity.

Particular symplectic morphisms are the *symplectic transvections* $\tau_{w,\lambda}$. For $w \in V$ and $\lambda \in \mathbb{R}$, $\tau_{w,\lambda} : V \to V$ is defined by

$$v \mapsto \lambda\omega\,(v,\, w)\, w.$$

EXERCISE 1.22. Show that

a) $\tau_{w,\lambda} \in Sp(V)$.

b) $Sp(V)$ is generated by the symplectic transvections (see JACOBSON [**Ja**], p. 373).

1.3. Subspaces of symplectic vector spaces

Let W be a k–dimensional linear subspace of the $2n$–dimensional symplectic space $(V,\, \omega)$. Then $k = \dim \phi(W)$ and $2l := \text{rank } \omega|_{\phi(W)}$ remain unchanged under any symplectic morphism from $Sp\,(V)$. We will show that these two integers, k and $2l$, classify the subspaces of V, in that they are the only two independent symplectic invariants for subspaces.

As in the proof of Proposition 1.3, let

$$W^{\perp} := \{v \in V;\ \omega(v, w) = 0 \quad \text{for all } w \in W\}.$$

$W^{\perp}$ is called the *skew* (or ω–)*orthogonal space of* W (and often simply called its *orthogonal*).

Remark 1.23. We have

$$\dim W^{\perp} = \dim V - \dim W = 2n - k.$$

Then the dimension formula from linear algebra (see, for example, LANG [**L1**], p. 95),

$$\dim V = \dim W + \dim W^{\circ}$$

for

$$W^{\circ} := \{\varphi \in V^{*},\ \varphi(w) = 0 \text{ for all } w \in W\},$$

can be here applied to show that (for an arbitrary non-degenerate bilinear form ω) $W^{\perp}$ can be identified with W° via ω^{b}.

Specific to the symplectic algebra is the appearance of the *radical* rad W of W of the form

$$\text{rad } W := W \cap W^{\perp}.$$

This is also given by

$$\text{rad } W = \{w \in W;\ i(w)\,\omega\big|_{W} = 0\}$$

and satisfies

$$\dim \text{ rad } W = \dim W - \text{ rank } \omega\big|_{W} = k - 2l.$$

Clearly, the structure of ω–orthogonal spaces has the following properties:

Remark 1.24.

a) $W^{\perp} \subset W'^{\perp}$ for $W' \subset W$,

b) $(W^{\perp})^{\perp} = W$,

c) $(W + W')^{\perp} = W^{\perp} \cap W'^{\perp}$,

d) $(W \cap W')^{\perp} = W^{\perp} + W'^{\perp}$.

A third space fixed by W is defined by

$${}^{u}W := W + W^{\perp}.$$

The usual dimension formula here says that

$$\dim {}^{u}W = \dim W + \dim W^{\perp} - \dim \text{ rad } W = 2n - (k - 2l).$$

From this space we can form yet more spaces, in particular

$$W^{red} := W/\text{rad } W,$$

the *symplectic space associated to* W. This satisfies dim $W^{red} = 2l$, and for a subspace $U \subset V$ with

$$W^{red} = \text{rad } W \oplus U,$$

$(U, \omega|_U)$ is symplectic and isomorphic to W^{red}. U is also ω–orthogonal to rad W. Therefore, we can also write $W = \text{rad } W \perp U$. Further, U' with $W^\perp = \text{rad } W \oplus U'$ is symplectic and isomorphic to the reduced space $(W^\perp)^{red} = W^\perp/\text{rad}\, W$.

The most important examples of subspaces of a symplectic space (V, ω) are the following.

DEFINITION 1.25.

1) A subspace Q of V with $\omega|_Q = 0$ is called an *isotropic subspace* of (V, ω).

2) A subspace $W \subset V$ with $\omega|_W$ non-degenerate is called a *symplectic subspace* of V.

3) A subspace $W \subset V$ with $W^\perp$ isotropic is called *coisotropic.*

4) A subspace $L \subset V$ which is both isotropic and coisotropic (thus with $L^\perp = L$) is called a *Lagrangian subspace.*

Before we discuss Lagrangian subspaces further, we prove the earlier statement that the dimension and rank are the only two independent symplectic invariants of a subspace. This result will be a corollary of the following otherwise useful discussion, which is here taken from ARTIN'S masterful parallel presentations of orthogonal and symplectic geometry in his *Geometric Algebra* ([**Ar**], pp. 114 f).

A two-dimensional symplectic space P is, according to Proposition 1.3, of the form

$$P = eK + e_*K \quad \text{with } \omega(e, e) = \omega(e_*, e_*) = 0,\ \omega(e, e_*) = 1.$$

P is also called a *hyperbolic plane* and (e, e_*) a *hyperbolic pair.* An orthogonal sum of hyperbolic planes is then called a *hyperbolic space*

$$H_{2r} = P_1 \perp \ldots \perp P_r.$$

By the same proposition, a symplectic space is always an orthogonal sum of hyperbolic planes and so is a hyperbolic space. An important step in proving our claim will be the following theorem, which says that every subspace of a symplectic space has a symplectic hull. On the way the notion of symplectic morphism will be extended to a morphism of those subspaces U which are compatible with the restriction of the symplectic form to U.

THEOREM 1.26. *Let V be symplectic and U a subspace, written in the form*

$$U = \text{rad } U \perp W.$$

Let $e_1, \dots, e_r$ be a basis of rad U*. Then there are elements $e_{1*}, \dots, e_{r*} \in V$ such that*

$$P_i = e_i K + e_{i*} K \quad \textit{is hyperbolic } (i = 1, \dots, r)$$

with

$$P_i \perp P_j \quad \textit{for } i \neq j, \textit{ and } \quad P_i \perp W \textit{ for all } i.$$

Then

$$V \supset \overline{U} := P_1 \perp \dots \perp P_r \perp W$$

is symplectic with $\overline{U} \supset U$.

An injective symplectic morphism ϕ from U into a symplectic space V' can be extended to a symplectic morphism $\overline{\phi} : \overline{U} \to V'$.

Proof. i) The proof of the first statement goes by induction. For $r = 0$ there is nothing to show. Now assume the claim is true for $r' < r$. Let

$$U_0 := \langle e_1, \dots, e_{r-1} \rangle \perp W.$$

Then e_r is orthogonal to U_0, $e_r \notin U_0$, and

$$\text{rad } U_0 = \text{rad } U_0^{\perp} = \langle e_1, \dots, e_{r-1} \rangle.$$

Consequently, we have $e_r \in U_0^{\perp}$ and $e_r \notin \text{rad } U_0^{\perp}$, and so there is an $a \in U_0^{\perp}$ for which $\omega (e_r, a) \neq 0$. The plane spanned by e_r and a is contained in $U_0^{\perp}$, and in this plane a can be changed to e_{r*} so that e_r, e_{r*} is a hyperbolic pair. It follows that

$$P_r := e_r K + e_{r*} K$$

is contained in $U_0^{\perp}$ and is orthogonal to U_0, and thus $U_0 \subset P_r^{\perp}$.

Since dim rad $U_0 = r - 1$, the induction hypothesis may be applied to

$$U_0 = \text{rad } U_0 \perp W \subset P_r^{\perp},$$

giving the existence of hyperbolic planes in U_0

$$P_i = \langle e_i, e_{i*} \rangle \subset P_r^{\perp}, \quad i = 1, \dots, r-1,$$

which are pairwise orthogonal and each orthogonal to W. Since all P_i are also orthogonal to P_r, we have that

$$\overline{U} := P_1 \perp \dots \perp P_r \perp W$$

is the desired symplectic space.

ii) Let ϕ be a symplectic morphism from U to V' and $e_i' := \phi (e_i)$ so that $W' := \phi (W)$. Then

$$\phi (U) = \langle e_1', \dots, e_r' \rangle \perp W'.$$

From i) there are elements $e'_{i*} \in V'$ so that for $P'_i = \langle e'_i, e'_{i*} \rangle$ we have

$$P'_i \perp P'_j \quad \text{for } i \neq j, \text{ and } P'_i \perp W', \quad i = 1, \ldots, r.$$

The rule $\overline{\phi}(e_{i*}) := e'_{i*}$ then gives the desired extension of ϕ. □

A consequence of this theorem is a special case of a theorem of Witt.

COROLLARY 1.27. *Let V and V' be isomorphic symplectic spaces, $U \subset V$ a subspace and ϕ an injective symplectic morphism from U to V'. ϕ can be extended to an isomorphism*

$$\overline{\phi} : V \to V'.$$

Proof. From the previous theorem, ϕ can be extended to a morphism $\overline{U} \to V'$. Thus we may assume that U is symplectic, and that $V = U \perp U^{\perp}$. Let $U' := \phi(U)$ and $U'^{\perp}$ be such that $V' = U' \perp U'^{\perp}$. It only remains to show that $U^{\perp}$ and $U'^{\perp}$ are symplectically isomorphic. But this is clear, since both spaces have the same dimension and no radical, and symplectic spaces of the same dimension are isomorphic. □

It is also now clear that dimension and rank of a subspace $U \subset V$ are the only symplectic invariants; given two subspaces U_i with the same rank and dimensional, we can as in the theorem decompose them into rad U_i and W_i, pair the bases of the radicals and W_i against each other, and then use the corollary to extend this map to a symplectic automorphism of V.

Besides the isotropic subspaces, the *Lagrangian subspaces* are the most important. Let's collect here some of the theory which can now be stated, given what has already been covered.

For a subspace $W \subset V$ with $\dim W = k$, we have

$$\begin{array}{llcll} W & \text{isotropic} & \Leftrightarrow & W \subset W^{\perp} & \Rightarrow k \leq n, \\ W & \text{coisotropic} & \Leftrightarrow & W \supset W^{\perp} & \Rightarrow k \geq n, \\ W & \text{Lagrange} & \Leftrightarrow & W = W^{\perp} & \Rightarrow k = n. \end{array}$$

A Lagrangian subspace L can thus be described as maximal isotropic. We then also have (with the identification $V \cong V^*$ via ω^b as in Section 1.1)

$$V = L \oplus L^{\perp} = V^*.$$

Remark 1.28. Let W be a coisotropic subspace of V. Then the image of $L \cap W$ under projection to W^{red} is a Lagrangian subspace.

EXERCISE 1.29. Prove this last remark. (See VAISMAN [**V**], p. 35.)

DEFINITION 1.30. Denote by $\mathcal{L}(V)$ the collection of Lagrangian subspaces $L \subset V$.

From the previous corollary, $Sp(V)$ acts transitively on $\mathcal{L}(V)$; that is, for any pair $L_1, L_2 \in \mathcal{L}(V)$, there is a $\phi \in Sp(V)$ with

$$\phi(L_1) = L_2.$$

In such a case, it is a general fact that $\mathcal{L}(V)$ has the structure of a *homogeneous space*. The concept of *homogeneous space* will later play a very important role (see Section 2.5). We will briefly make its acquaintance here with a special example.

Denote by G_L the *isotropy group* of $L \in \mathcal{L}(V)$; that is,

$$G_L := \{\phi \in Sp(V),\ \phi(L) = L\}.$$

Let $\underline{e} = (e_i, e_{i*})_{i=1,\dots,n}$ be an *L–related* symplectic basis of V; that is, $\langle e_1, \dots, e_n\rangle = L$ and $\langle e_{1*}, \dots, e_{n*}\rangle = L^{\perp}$. Then, relative to this basis, $V \cong K^{2n}$, $Sp(V) \cong Sp_n(K)$, $L \cong L_0 := \langle e_1^\circ, \dots, e_n^\circ\rangle$, where (e_i°) are the canonical unit vectors in K^{2n}, and

$$G_L \cong G_{L_0} = \left\{\begin{pmatrix} A & B \\ 0 & D \end{pmatrix};\ {}^tBD = {}^tDB,\ {}^tAD = E_n\right\}.$$

This is because $M(L_0) = L_0$, and thus

$$\begin{pmatrix} A & B \\ C & D \end{pmatrix}\begin{pmatrix} e' \\ 0 \end{pmatrix} \in L_0$$

forces $C = 0$, so that from Remark 1.13 about the general form of symplectic matrices we get the claimed transitive action.

From this we get a bijection between $\mathcal{L}(V)$ and the set of cosets of G_{L_0} in $Sp_n(K)$; thus

$$Sp_n(K)/G_{L_0} \cong \mathcal{L}(V),\ M \longmapsto M(L_0),$$

because, from

$$M(L_0) = M'(L_0),$$

it follows that

$$M^{-1}M' \in G_{L_0}.$$

In a similar vein, we can describe another family of interesting subspaces of a given symplectic space. Fix $\mathcal{T}(L)$, for a given Lagrangian subspace $L \subset V$, to be the collection of all Lagrangian subspaces L' transverse to L, so that

$$\mathcal{T}(L) := \{L' \in \mathcal{L}(V),\ L \oplus L' = V\}.$$

The proofs already given show that G_L operates transitively on $\mathcal{T}(L)$. Take (e_i) to be a basis of L and (e_{i*}) a basis of L' (so that both together form a symplectic basis of V). Then, relative to this basis, the isotropy group

$G_{L,L'} \subset G_L \subset Sp\,V$ that fixes L and L' can be identified with the group of matrices

$$\begin{pmatrix} A & 0 \\ 0 & {}^tA^{-1} \end{pmatrix}, \quad A \in GL_n(K),$$

which is naturally isomorphic to $GL_n(K)$. We then get that

$$\mathcal{T}\,(L) \cong G_L/GL_n(K).$$

In VAISMAN [**V**], pp. 36–38, the following statement and a few of its consequences are discussed.

THEOREM 1.31. *$\mathcal{T}\,(L)$ has a natural structure as an affine space over K of dimension $n\,(n+1)/2$.*

An affine space is a triple $(\mathcal{A}, V, \pi)$, where $\mathcal{A}$ is a set, V is a K vector space and $\pi : \mathcal{A} \times \mathcal{A} \to V$ is a morphism, such that $\pi|_{\{a_0\}\times\mathcal{A}}$, for a fixed point $a_0 \in \mathcal{A}$, is a bijection and for all $a, b, c \in \mathcal{A}$ we have

$$\pi(a,\, b) + \pi(b,\, c) = \pi(a,\, c).$$

This definition can easily be brought into agreement with the definitions which often appear in elementary textbooks. Of the proof which is contained in VAISMAN [**V**], we give only a sketch:

By choosing the earlier required basis, the matrix of a morphism which carries $L' \in \mathcal{T}\,(L)$ to $L'' \in \mathcal{T}\,(L)$ has the form

$$\begin{pmatrix} A & B \\ 0 & D \end{pmatrix} \in G_{L_0},$$

which then modulo $G_{L,L'} \cong GL_n(K)$ has the form

$$\begin{pmatrix} E_n & X \\ 0 & E_n \end{pmatrix} \quad \text{with } X = {}^tX.$$

This shows already that $\mathcal{T}\,(L)$ relative to a fixed basis of V is in bijection to a set of symmetric $n \times n$ matrices, and therefore can be seen as an $\big(n\,(n+1)/2\big)$–dimensional affine subspace of K^{n^2}. This observation can be made basis–independent, in that the symmetric matrix X can be defined as a coordinate–independent quadratic form q on the dual space L^*. The following remarks can then be established:

Remark 1.32. For any pair (L, L') of Lagrangian subspaces of (V, ω), there exists a mutually transverse Lagrangian space L''.

Remark 1.33. Let L, L', L'' be Lagrangian subspaces of $(V,\, \omega)$, and let $L \cap L' = L \cap L''$. Then there exists a (not unique) symplectic transformation of V which fixes every vector of L and carries L' to L''.

EXERCISE 1.34. The reader would benefit from supplying proofs for these remarks (which in any case can be found in VAISMAN [**V**]).

1.4. Complex structures of real symplectic spaces

Up to now, we have considered only $\mathbb{R}^{2n}$ equipped with

i) the *canonical Euclidian structure* with the form

$$s(v,\, w) := (v,\, w) = {}^t v w = \sum_{j=1}^{2n} v_j w_j \qquad \text{for } v, w \in \mathbb{R}^{2n} \text{ (as columns)},$$

ii) the *canonical symplectic structure* with the form

$$\omega(v,\, w) = {}^t v J_n w = \sum_{i=1}^{n} (v_i\, w_{n+i} - v_{n+i}\, w_i).$$

Now, as $\mathbb{R}$ vector spaces, $\mathbb{R}^{2n} \simeq \mathbb{C}^n$. With the identification

$$v = \begin{pmatrix} x \\ y \end{pmatrix} \leftrightarrow z = x + iy$$

the operation $z \mapsto iz$ corresponds to the operation $v \mapsto -J_n v = \begin{pmatrix} -y \\ x \end{pmatrix}$.
This operation,

$$J_n : \mathbb{R}^{2n} \to \mathbb{R}^{2n} \quad \text{with } J_n^2 = -\text{ id},$$

supplies $\mathbb{R}^{2n}$ with a complex structure. It is then natural ask in how many ways $\mathbb{R}^{2n}$ or, indeed, an arbitrary $\mathbb{R}$ vector space, can be supplied with a complex structure. This question will be considered further below. For now, we give a brief survey.

The invertible linear morphisms of $\mathbb{R}^{2n}$, that is, the invertible matrices M, that

i) preserve the *canonical scalar product*, s, form the *orthogonal group* $O(2n)$;

ii) preserve the *symplectic standard-form*, ω, form the *symplectic group* $Sp_n(\mathbb{R})$;

iii) preserve the *complex structure* J, that is, with $MJ = JM$, are of the form

$$M = \begin{pmatrix} X & -Y \\ Y & X \end{pmatrix},$$

and form the *general linear group* $GL_n(\mathbb{C})$ via $M \mapsto X + iY$.

It can be shown that

$$O(2n) \cap Sp_n(\mathbb{R}) = Sp_n(\mathbb{R}) \cap GL_n(\mathbb{C}) = GL_n(\mathbb{C}) \cap O(2n) = U(n).$$

The unitary group $U(n)$ then preserves the hermitian scalar product

$$h(v, w) = s(v, w) - i\,\omega(v, w).$$

As a consequence of earlier material, we have, for the space of all Lagrangian subspaces of $\mathbb{R}^{2n}$,

$$\mathcal{L}(\mathbb{R}^{2n}) \simeq U(n)/O(n),$$

and the space of all *positively compatible* complex structures J on $\mathbb{R}^{2n}$ is in bijection with

$$\mathfrak{H}_n = \{Z \in M_n(\mathbb{C}); {}^tZ = Z,\ \mathrm{Im}\, Z > 0\} = Sp_n(\mathbb{R})/U(n).$$

All of this will now be discussed in greater generality (following VAISMAN ([**V**], pp. 40 ff.).

DEFINITION 1.35. Let V be an $\mathbb{R}$ vector space. $J \in \mathrm{Aut}\, V$ is called a *complex structure* on V if and only if

$$J^2 = -\mathrm{id}_V.$$

In the case that V is symplectic with the form ω, then we call the complex structure J *compatible* with ω if

$$\omega(Jv, Jw) = \omega(v, w) \text{ for all } v, w \in V.$$

Slightly changing the notation as in the previous example, (V, J) with an arbitrary J can be made into a $\mathbb{C}$ vector space via

$$\sqrt{-1}v := Jv.$$

Further, J can be extended linearily to the *complexification*

$$V_c := V \otimes_{\mathbb{R}} \mathbb{C}.$$

Then J has the eigenvalue $\pm\sqrt{-1}$, and from $Jw = \lambda w$ we deduce that

$$-w = J^2 w = \lambda J w = \lambda^2 w.$$

The eigenspace of $\lambda_{1,2} = \pm\sqrt{-1}$ is n–dimensional, and is given by

$$V_c^+ := \{v - \sqrt{-1}\, Jv\ v \in V\}, \qquad \overline{V}_c^+ := \{v + \sqrt{-1}\, Jv,\ v \in V\}.$$

We then have that

$$V_c = V_c^+ \oplus \overline{V}_c^+,$$

and

$$v \mapsto v - \sqrt{-1}\, Jv$$

defines a $\mathbb{C}$ vector space isomorphism between (V, J) and $(V_c^+, \sqrt{-1})$.

In the case that J is a complex structure compatible with the symplectic form ω, we have

$$g(v, w) := \omega(v, Jw) \qquad \text{for } v, w \in V.$$

From the compatibility of J and ω we have also

$$g(Jv, w) = \omega(v, w),$$

and from $J^2 = -1$ and the skew–symmetry of ω we also get

$$g(v, w) = g(w, v)$$

and

$$g(Jv, Jw) = g(v, w).$$

Therefore g is a symmetric bilinear form, and, like ω, is also non–degenerate. g will be called an *ω–compatible pseudohermitian metric.* When $g(v, v) \geq 0$ for all $v \in V$, we call g a *hermitian*[1] *metric*, J a *positive compatible* complex structure and the triple (V, ω, J) a *Kähler vector space.*

THEOREM 1.36. *Every real symplectic vector space (V, ω) can be given a compatible positive complex structure J and a hermitian structure g. Any two such structures J_0 und J_1 are homotopic in the following sense: there is a differentiable family J_t, $0 \leq t \leq 1$, of positive compatible complex structures on V defining a path from J_0 with J_1.*

Remark 1.37. An analogous statement holds for Hilbert spaces having a skew–symmetric weakly nondegenerate bilinear form ω. The following reasoning for the general case reduces the proof of ABRAHAM–MARSDEN ([**AM**], p. 173) to our simpler case.

Proof. Let γ be a Euclidean scalar product on V, and let $A : V \to V$ be defined by

$$\gamma(Av, w) = \omega(v, w) \qquad \text{for all } v, w \in V.$$

Since ω is skew–symmetric, we have

$$\gamma(Av, w) = -\omega(w, v) = -\gamma(v, Aw),$$

and so

$$\gamma(A^2v, w) = -\gamma(Av, Aw) = \gamma(v, A^2w);$$

this is to say that A^2 is self–adjoint and, since

$$\gamma(A^2v, v) = -\gamma(Av, Av) \leq 0$$

is negative, must have negative, but not necessarily distinct, eigenvalues $-\lambda_j^2$, $\lambda_j > 0$ $(j = 1, \ldots, 2n)$. V then has a γ–orthonormal–basis $\underline{a}$ of eigenvectors for A^2, $a_1, \ldots, a_{2n}$. Let $B \in \text{Aut } V$ with

$$M_{\underline{a}}(B) = \begin{pmatrix} \lambda_1 & & \\ & \ddots & \\ & & \lambda_{2n} \end{pmatrix}.$$

Then B is the unique self–adjoint positive operator with

$$B^2 = -A^2.$$

[1]This notation will later be clarified: it will be shown that g can be extended to a hermitian metric on the complexification of V.

For $J := AB^{-1}$ we have, when we consider the eigenspaces of A and B,

$$J = B^{-1}A,$$

and thus

$$J^2 = AB^{-1}.\, B^{-1}A = -E.$$

This J is compatible with ω, and so

$$\begin{aligned}
\omega(Jv,\, Jw) &= \omega(AB^{-1}v,\, AB^{-1}w)\\
&= \gamma(A^2B^{-1}v,\, AB^{-1}w) = -\gamma(AB^{-1}v,\, A^2B^{-1}w)\\
&= -\omega(B^{-1}v,\, A^2B^{-1}w) = \omega(B^{-1}v,\, Bw)\\
&= \omega(BB^{-1}v,\, w) = \omega(v,\, w).
\end{aligned}$$

for g with

$$(*) \qquad g\,(v,\, w) := \omega\,(v,\, Jw),$$

and we have

$$\begin{aligned}
g(v,\, v) &= \omega(v,\, Jv) = \gamma(Av,\, Jv) = \gamma(v,\, AJv)\\
&= \gamma(v,\, Bv) \geq 0
\end{aligned}$$

by the construction of B. This J satisfies the requirements of the theorem. Since it is dependent on the chosen scalar product γ, we write $J = J_\gamma$. It can be seen that every positive compatible complex structure J arises in this manner from some such γ (one need just note that $J = J_g$ with g from $(*)$).

The last statement of the theorem is now easy. Let J_0 and J_1 be given; then they are of the form J_{γ_0} and J_{γ_1}, and can be carried from one to the other via the family J_{γ_t}, with

$$\gamma_t := t\gamma_0 + (1-t)\gamma_1 \quad (0 \leq t \leq 1).$$

□

In the *standard case*,

$$\begin{aligned}
V &= \mathbb{R}^{2n},\\
\text{with} \quad \omega\,(v,\, w) &= {}^t v J_n w,\\
J_n &= \begin{pmatrix} 0 & 1 \\ -1 & 0 \end{pmatrix},\\
\text{and} \quad \gamma\,(v,\, w) &= {}^t v w,
\end{aligned}$$

we clearly have $A = -J_n$. So $A^2 = -E$, $B = E$, and so $J = -J_n$ gives the positive complex structure, and $g\,(v,\, w) = {}^t v w$.

The scalar product described above is called the *hermitian metric*, since g induces a *hermitian bilinear form* g_c on the complexification V_c of V by

$$g_c(\sqrt{-1}\,v,\, w) = -g_c(v,\, \sqrt{-1}\,w) = \sqrt{-1}\,g\,(v,\, w) \quad \text{for } v,\, w \in V.$$

Then there is the restriction of g_c to V_c^+

$$g_c(v - \sqrt{-1}\,Jv,\, w - \sqrt{-1}\,Jw) = 2\left[g\,(v,\, w) - \sqrt{-1}\,\omega(v,\, w)\right],$$

and then

$$h\,(v,\, w) := g\,(v,\, w) - \sqrt{-1}\,\omega(v,\, w) \text{ for } v,\, w \in V$$

is a usual hermitian metric on $(V,\, J)$ as $\mathbb{C}$ vector space. This immediately gives us

Remark 1.38. The map

$$V \ni v \mapsto \frac{1}{\sqrt{2}}(v - \sqrt{-1}\,Jv) \in V_c^+$$

is an isomorphism of the hermitian spaces $(V,\, J,\, h)$ and $(V_c^+,\, g_c)$.

From linear algebra $(V,\, J,\, h)$ has an orthonormal basis relative to h, thus a $\mathbb{C}$–basis (e_j), $j = 1, \dots, n$, with

$$h\,(e_j,\, e_k) = \delta_{jk}.$$

Since $h = g - \sqrt{-1}\,\omega$ and $g\,(v,\, w) = g\,(Jv,\, Jw) = \omega\,(v,\, Jw)$, this is equivalent to

$$g(e_j,\, e_k) = g(Je_j,\, Je_k) = \delta_{jk}, \quad g(e_j,\, Je_k) = 0.$$

From this we may conclude that $(e_j,\, Je_j;\ j = 1, \dots, n)$ is a real g orthonormal basis of V and, with $e_{j*} = e_{n+j} = Je_j$, the family $(e_j,\, e_{j*})$, $j = 1, \dots, n$, is a symplectic basis of V.

In the other direction, if $(e_j,\, Je_j)$ is a *real unitary basis* of $(V,\, \omega,\, J)$, that is, a g orthonormal basis and at the same time a symplectic basis, then (e_j) is also an h–basis, and we have a corresponding $\mathbb{C}$–basis of V_c^+ given by

$$e_j' = (e_j - \sqrt{-1}\,Je_j)/\sqrt{2}\,, \quad j = 1, \dots, n.$$

This is called a *complex unitary* basis.

The automorphisms of the structure $(V,\, \omega,\, J)$ which leave the symplectic form ω, as well as the metric g (or, equivalently, those which commute with J), fixed are called the *unitary transformations* and generate the *unitary group* $U(V,\, J)$. These automorphisms are also characterized by carrying unitary bases into unitary bases. Other equivalent descriptions are given as the complex–linear transformations of $(V,\, J)$ which fix the metric h, or (see the last remark) as the group of complex linear transformations of V_c^+ which fix the metric g_c. By fixing a basis one may derive the relationships between matrix groups as described at the beginning of this section: The choice of a unitary basis $(e_j;\ j = 1, \dots, n)$ allows one to identify $(V,\, J,\, h)$ as well as (V_c^+, g_c) with $\mathbb{C}^n$, where h in the complex coordinates (z_j) relative to (e_j) is the canonical hermitian metric

$$h(z, z') = \sum_{j=1}^{n} z_j \overline{z}'_j.$$

Here we then have that

$$U(V, J) \simeq U(n) = \{U \in GL_n(\mathbb{C});\ U^t\overline{U} = E_n\}.$$

On the other hand, for the Euclidian vector space (V, g), $O(V, g)$ is the orthogonal group of linear isomorphisms which fix g. With the help of a g orthonormal basis this group is identified with $O(2n)$. The above discussion can be summarized in the following proposition.

THEOREM 1.39. *Let (V, ω) be a real symplectic space and J a positive compatible complex structure. Then*

$$U(V, J) = Sp(V) \cap O(V, g).$$

In particular, for $V = \mathbb{R}^{2n}$ with

- the canonical basis $\underline{e} = (e_i, e_{i*})$ and the coordinates $\begin{pmatrix} v_i \\ v_{i*} \end{pmatrix}$,
- the symplectic standard form
$$\omega(v, w) = {}^t v J w = \sum_{i=1}^{n} (v_i\, w_{i*} - v_{i*}\, w_i),$$
- the canonical scalar product
$$g(v, w) = {}^t v w = \sum_{i=1}^{n} (v_i w_i + v_{i*}\, w_{i*}),$$
- the complex structure defined by $Je_i = e_{i*}$, $Je_{i*} = -e_i$,

then the complex coordinates $z_i = v_i + \sqrt{-1}\, v_{i*}$ $(i = 1, \ldots, n)$ satisfy

$$U(n) = Sp_n(\mathbb{R}) \cap O(2n).$$

This leads to the following alternative formulation of the description given in Section 1.3 of the space $\mathcal{L}(V)$ of Lagrangian subspaces $L \subset V$ in the case that V has a compatible positive complex struture J. This arises from the fact that in the above introduced notation, any g orthonormal basis $(e_i)_{i=1,\ldots,n}$ of L $(e_i,\, Je_i)$ gives rise to a real unitary basis of V. Each such basis is called an *L–related unitary basis.* For $L' \in \mathcal{L}(V)$ with L'–related basis (e_i', Je_i') there is a unitary transformation which takes (e_i, Je_i) to (e_i', Je_i'), and so L to L'. The unitary group $U(V, J)$ operates transitively on $\mathcal{L}(V)$, and the isotropy group which fixes an L is the group which carries L–related bases to themselves and is thus isomorphic to the orthogonal group $O(L, g)$. Thus we have shown

THEOREM 1.40. *We have that*

$$\mathcal{L}(V) \simeq U(V, J)/O(L, g) \simeq U(n)/O(n).$$

Now we will fulfill the promise made at the beginning of this section; that is, we describe the collection

$$\mathcal{J} = \mathcal{J}(V, \omega)$$

of all compatible positive complex structures J on real symplectic space (V, ω), and so the set of possible ways (V, ω) can be made into a Kähler vector space. From this it will follow that $\mathcal{J}$ can be identified with the Siegel upper half plane

$$\mathfrak{H}_n \simeq Sp_n(\mathbb{R})/U(n).$$

From Remark 1.37, for a fixed $J \in \mathcal{J}$, the space (V, J) can be identified with a subspace V_c^+ of the complexification of V_c. Here V_c^+ is a Lagrangian subspace of V_c, where the symplectic structure ω of V is linearly extended to V_c. This is because we have

$$\begin{aligned}&\omega(v - \sqrt{-1}\,Jv,\, w - \sqrt{-1}\,Jw)\\ &\qquad = \omega(v,\, w) - \omega(Jv,\, Jw) - \sqrt{-1}\,\big(\omega(Jv,\, w) + \omega(v,\, Jw)\big) = 0.\end{aligned}$$

Furthermore, since $g(v,\, w) := \omega(v,\, Jw)$,

$$-\sqrt{-1}\,\omega(v,\, \overline{v}) = 2g(a,\, a) > 0 \ \text{ for } 0 \neq v := a - \sqrt{-1}\,Ja \in V_c^+,$$

and, additionally, $Jv = \sqrt{-1}(v_1 - v_2)$ for

$$V_c = V_c^+ \oplus \overline{V}_c^+ \ni v = v_1 + v_2, \quad v_1 \in V_c^+,\, v_2 \in \overline{V}_c^+.$$

Then the Lagrangian subspace F of $(V_c,\, \omega)$ is called *positive* if

$$-\sqrt{-1}\,\omega(v, \overline{v}) > 0 \qquad \text{for all } 0 \neq v \in F.$$

Remark 1.41. There is a natural bijection between $\mathcal{J} = \mathcal{J}(V,\, \omega)$ and the collection $\mathcal{L}_+ = \mathcal{L}_+(V_c)$ of positive Lagrangian subspaces of $(V_c,\, \omega)$.

Proof. i) The mapping $\mathcal{J} \to \mathcal{L}_+$ is defined from the map $J \mapsto V_c^+$.

ii) For $F \in \mathcal{L}_+$ $\overline{F}$ is also a Lagrangian subspace of V_c, since ω is real. From the positivity of F, it follows that $F \cap \overline{F} = \{0\}$ and so $V_c = F \oplus \overline{F}$. Now we can define $J : V_c \to V_c$ by

$$J(v_1 + \overline{v}_2) = \sqrt{-1}\,(v_1 - \overline{v}_2) \text{ for } v_1, v_2 \in F.$$

Then $J^2 = -\mathrm{id}$, $\omega(Jv, Jw) = \omega(v, w)$, for all $v, w \in V_c$, and $J(V) = V$, and so for $v \in V_c$ we have the bijection

$$v \in V \Leftrightarrow v = \overline{v}.$$

Finally, we have

$$\omega(v, Jv) = -2\sqrt{-1}\,\omega(v_1, \overline{v}_1) > 0 \quad \text{for } 0 \neq v = v_1 + \overline{v}_1 \in V,$$

and so $J \in \mathcal{J}(V, \omega)$.

iii) The morphisms $\mathcal{J} \to \mathcal{L}_+$ from i) and $\mathcal{L}_+ \to \mathcal{J}$ from ii) are clearly inverse to one another. □

A Lagrangian subspace L_c of a complex symplectic space V_c is called a *real Lagrangian subspace*, if it is the complexification of a Lagrangian subspace $L \subset V$; that is, $L_c = L \otimes_{\mathbb{R}} \mathbb{C}$. This is satisfied precisely when L_c is carried to itself by complex conjugation, that is, when $L_c = \overline{L}_c$. In V_c, all real Lagrangian subspaces L_c are transversal to every $F \in \mathcal{L}_+$. This is because for

$$0 \neq v = a + \sqrt{-1}\,b \in L_c \cap F \quad \text{with } a, b \in L$$

we have

$$0 < -\sqrt{-1}\,\omega(v, \overline{v}) = -2\omega(a, b) = 0,$$

and so also the converse.

The remainder of the treatment in Vaisman [**V**] is based on the description of the space $\mathcal{T}(L)$ of Lagrangian subspaces transverse to a fixed Lagrangian subspace as an affine $n(n+1)/2$–dimension space, as given in Theorem 1.31. Since this description was not fully given here, we can only give a sketch of Vaisman's treatment, but enough to see the idea:

Let L_c be a fixed real Lagrangian space. Then, as described above, $\mathcal{L}_+ = \mathcal{L}_+(V_c) \subset \mathcal{T}(L_c)$. That $\mathcal{T}(L_c)$ is an affine space means that for each pair $F \in \mathcal{L}_+$ and $L'_c \in \mathcal{T}(L_c)$ with L'_c real and transversal to L_c (thus $L_c \oplus L'_c = V$) there is a symplectic transformation ϕ which fixes every point of L_c and carries L'_c into F. If $\underline{e} = (e_i, e_{i*})$ $(e_i \in L,\ e_{i*} \in L')$ is a basis of V, respectively V_c, appropriate to L, then ϕ (see Theorem 1.31), relative to this basis, is written as a matrix

$$\begin{pmatrix} E_n & Z \\ 0 & E_n \end{pmatrix} \quad \text{with } Z \in M_n(\mathbb{C}),\ Z = {}^tZ.$$

Then $\det Z \neq 0$, since F is also transversal to L'_c. The positivity of F sets yet another condition on Z. And thus ϕ can be written, relative to $\underline{e}$, as

$$((e_i), (e_{i*}))\begin{pmatrix} E & Z \\ 0 & E \end{pmatrix} = ((e_i), (e_{i*}) + (e_i)Z),$$

and so

$$\phi(e_i) = e_i, \quad \phi(e_{i*}) = e_{i*} + \sum_{j=1}^{n} e_j Z_{ji}, \ (Z_{ij}) = Z.$$

In fact, $\phi(e_{i*})$ is a basis of F, and a small calculation shows that F is positive, and so

$$-\sqrt{-1}\,\omega(v, \overline{v}) > 0 \quad \text{for } 0 \neq v \in F$$

exactly when

$$\operatorname{Im} Z = \frac{1}{2\sqrt{-1}}(Z - \overline{Z})$$

is a positive definite matrix, which can also be written simply as $\operatorname{Im} Z > 0$. This shows that, with a choice of a basis $\underline{e}$, the mapping $F \mapsto Z$ gives a mapping from $\mathcal{L}_+$ into the so-called *Siegel upper half plane*

$$\mathfrak{H}_n = \{Z \in M_n(\mathbb{C}),\ {}^tZ = Z,\ \operatorname{Im} Z > 0\}.$$

With a bit more care and the help of Remark 1.41, one can show

THEOREM 1.42. *For a real $2n$–dimensional symplectic space (V, ω), the set $\mathcal{J}(V, \omega)$ of positive compatible complex structures on V can be identified with the Siegel upper half plane $\mathfrak{H}_n$.*

The identification of the theorem depends on the symplectic basis $\underline{e}$ of V. Should the symplectic basis be changed to $\underline{\tilde{e}}$, then, since

$$(*) \qquad (e, e_*) = (\tilde{e}, \tilde{e}_*)\begin{pmatrix} A & B \\ C & D \end{pmatrix} \quad \text{with} \quad \begin{pmatrix} A & B \\ C & D \end{pmatrix} \in Sp_n(\mathbb{R}),$$

there is a matrix $\widetilde{Z}$ associated to F relative to $\underline{\tilde{e}}$ with the property that F has the bases $e_* + eZ$ as well as $\tilde{e}_* + \tilde{e}\widetilde{Z}$. So there is a $\Lambda \in GL_n(\mathbb{C})$ with

$$e_* + eZ = (\tilde{e_*} + \tilde{e}\widetilde{Z})\Lambda.$$

Applying the transformation formula $(*)$, we get

$$\tilde{e}B + \widetilde{e}_*D + (\tilde{e}A + \widetilde{e}_*C)Z = \widetilde{e}_*\Lambda + \tilde{e}\widetilde{Z}\Lambda,$$

and thus

$$\Lambda = CZ + D \quad \text{and} \quad \widetilde{Z} = (AZ + B)(CZ + D)^{-1}.$$

From complex function theory, the mapping

$$Z \mapsto \widetilde{Z} = (AZ + B)(CZ + D)^{-1} =: \begin{pmatrix} A & B \\ C & D \end{pmatrix}\langle Z \rangle$$

is recognizable as a complex–analytic automorphism of $\mathfrak{H}_n$.

We have just seen that $Sp_n(\mathbb{R})$ operates on $\mathfrak{H}_n$, and previously we have established the bijection $\mathcal{L}(V) \simeq Sp_n(\mathbb{R})/G_{L_0}$. Therefore it is not so surprising that $\mathfrak{H}_n$, and hence $\mathcal{J}(V, \omega)$, can be described as a homogeneous space. It turns out that

$$\mathcal{J}(V, \omega) \simeq \mathfrak{H}_n \simeq Sp_n(\mathbb{R})/U(n).$$

Thus the operation $\phi \in Sp(V)$ can be extended to an operation on V_c which is also called ϕ. This ϕ then also operates on $\mathcal{L}_+$, and gives a *transitive* operation of $Sp(V)$ on $\mathcal{L}_+$. This is because the equality

$$(*), \qquad h(v, w) := -\sqrt{-1}\,\omega(v, \overline{w}) \quad \text{for } v, w \in F \text{ from } \mathcal{L}_+$$

allows one to define a hermitian metric h on F, and for $(f_i)_{i=1,\dots,n}$, an orthonormal basis of F, we have that $(f_i, -\sqrt{-1}\,\overline{f}_i)_{i=1,\dots,n}$ is then a symplectic basis of V_c. Analogously, for F' from $\mathcal{L}_+$ with h orthonormal basis (f_i'), $(f_i', -\sqrt{-1}\,\overline{f}_i')_{i=1,\dots,n}$ is a symplectic basis of V_c. Then there is a $\phi \in Sp(V_c)$ with

$$\phi(f_i) = f_i', \qquad \phi(\bar{f}_i) = \bar{f}_i'.$$

This ϕ maps F to F' and commutes with complex conjugation, and therefore is in $Sp(V)$. Thus, the transitivity of the operation of $Sp(V)$ on $\mathcal{L}_+$ is demonstrated. The isotropy group of F in $Sp(V)$ is, because of $(*)$, the unitary group $U(F, h)$. And so we have shown

THEOREM 1.43.

$$\mathcal{J}(V, \omega) \simeq \mathcal{L}_+(V) \simeq \mathfrak{H}_n \simeq Sp(V)/U(F, h) \simeq Sp_n(\mathbb{R})/U(n).$$

EXERCISE 1.44. The reader is recommended to show directly (that is, independently of what has come before) that:

a) $G = SL_2(\mathbb{R})$ operates on $\mathfrak{H}_1 = \{\tau = x + iy \in \mathbb{C},\ y > 0\}$ by

$$(M, \tau) \mapsto M(\tau) := \frac{a\tau + b}{c\tau + d} \quad \text{for } M = \begin{pmatrix} a & b \\ c & d \end{pmatrix} \in SL_2(\mathbb{R}) \text{ and } \tau \in \mathfrak{H}_1.$$

b) The following isomorphism holds:

$$\mathfrak{H}_1 \cong SL_2(\mathbb{R})/SO(2) \cong Sp_1(\mathbb{R})/U(1).$$

c) The *Poincaré metric*

$$ds^2 = \frac{dx^2 + dy^2}{y^2}$$

is $SL_2(\mathbb{R})$-invariant.

The Siegel half space $\mathfrak{H}_n$ has many uses in the area of moduli problems of Abelian varieties. It can be given the structure of a complex manifold of dimension $n(n+1)/2$ (see SATAKE [**Sa**], p. 78). The study of the geometry of these manifolds and the holomorphic, as well as the meromorphic, functions on $\mathfrak{H}_n$ with known invariant or covariant properties under the operation of the group $Sp_n(\mathbb{Z})$ or its subgroups was initiated by SIEGEL (see his *Symplectic Geometry* [**Si1**] or *Topics in Complex Function Theory* [**Si2**]); for some time it was exactly these topics introduced by Siegel that formed the subject of symplectic geometry. One may find a new and particularly nice treatment of these traditional topics in MUMFORD: *Tata Lectures on Theta, Vol. 2* ([**Mu1**]). Nowadays, however, symplectic geometry refers to a much broader range of topics, which we will consider in the next chapters.

Chapter 2

Symplectic Manifolds

Symplectic geometry arises from the globalization of the symplectic algebra considered in the previous chapter. The central concept is that of a symplectic manifold. We begin, in this chapter, by defining these, and continue by studying some of their properties and by giving several examples. A good reference for this material is the second chapter of AEBISCHER *et al.* [**Ae**] and Section 3.2 of ABRAHAM-MARSDEN [**AM**].

2.1. Symplectic manifolds and their morphisms

In what follows we will assume that M is a *smooth manifold* of dimension p, that is, a C^∞–manifold in the sense of Section A.1, and assume, as well, that it is *real*, unless something is said to the contrary.

DEFINITION 2.1. M is called a *symplectic manifold*, if there is defined on M a closed nondegenerate 2–form ω; that is, an $\omega \in \Omega^2(M)$ such that

i) $d\omega = 0$,

ii) on each tangent space T_mM, $m \in M$, if

$$\omega_m(X, Y) = 0 \quad \text{for all } Y \in T_mM,$$

then $X = 0$.

The assumptions on ω say that its restriction to each $m \in M$ makes the tangent space T_mM into a symplectic vector space. Thus it is already clear that the dimension p of M is even; thus $p = 2n$. In the next section, it will be shown that all symplectic manifolds of the same dimension are *locally the same*. This is in sharp contrast to the situation in Riemannian geometry, and indicates that symplectic geometry is essentially a global theory.

However, not every even–dimensional manifold has a symplectic structure. In AEBISCHER *et al.* ([**Ae**], p.3) is given the example of $M = S^4$; however, this uses a cohomological result (see Appendix C), which we should not yet go into.

Given two symplectic manifolds (M, ω) and (M', ω'), let $F : M \to M'$ be a smooth map, that is, differentiable in the sense of Section A.1.

DEFINITION 2.2. The map F is called *symplectic*, or a *morphism of symplectic manifolds*, so long as

$$F^*\omega' = \omega.$$

Given a symplectic diffeomorphism F, F^{-1} is also *symplectic*, and F is called a *symplectomorphism*. $Sp(M)$ denotes the *group of symplectomorphisms* from M to itself.

2.2. Darboux's theorem

Darboux's theorem has, in its simplest form, the following formulation. To every point m of a symplectic manifold (M, ω) of dimension $2n$, there correspond an open neighborhood U of m and a smooth map

$$F : U \to \mathbb{R}^{2n} \quad \text{with } F^*\omega_0 = \omega|_U,$$

where ω_0 is the standard symplectic form on $\mathbb{R}^{2n}$. It follows immediately that for an appropriate choice of *symplectic coordinates* $x = (q, p)$, $p = (p_1, \dots, p_n)$ and $q = (q_1, \dots, q_n)$, ω can be written on U in the form

$$\omega = dq \wedge dp = \sum_{i=1}^{n} dq_i \wedge dp_i.$$

In AEBISCHER *et al.* ([**Ae**], p. 17) a proof is given which follows that of GUILLEMIN-STERNBERG ([**GS**], pp. 156ff). We will use this proof here and supply somewhat more detail. It does go back (as the proof in [**AM**], p. 175) to MOSER (1965) and WEINSTEIN (1977). As a preliminary we give a standard result on the behavior of parameter–dependent differential forms under transformations, which will be useful in other places. Although the details are somewhat tedious, its proof will provide a good exercise in the calculus of differential forms described in Section A.4.

LEMMA 2.3. *Let M and M' be smooth manifolds and $F_t : M \to M'$, $t \in \mathbb{R}$, a smooth 1–parameter family of maps. Let X_t denote the* tangent field of M' along F_t*; that is, X_t is the map*

$$\begin{array}{ccl} M & \longrightarrow & TM', \\ m & \longmapsto & (m', X_t(m)), \quad m' = F_t(m), \end{array}$$

where $X_t(m)$ is the tangent vector at $m' = F_t(m) \in M'$ to the curve $s \mapsto F_s(m)$. Let $(\sigma_t)_{t\in\mathbb{R}}$ be a 1–parameter family of differential forms on M'. Then we have

$$\begin{aligned}\frac{d}{dt}(F_t^*\sigma_t) &= F_t^*\left(\frac{d\sigma_t}{dt} + i(X_t)d\sigma_t\right) + d(F_t^*(i(X_t)\sigma_t))\\ &= F_t^*\left(\frac{d\sigma_t}{dt} + i(X_t)d\sigma_t + d(i(X_t)\sigma_t)\right),\end{aligned}$$

so long as all the maps F_t are diffeomorphisms from M to M' and, therefore, X_t gives a vector field on M'.

The derivative $\frac{d}{dt}$ applied to a one–parameter (t) differential simply means the differentiation of the coefficients with respect to that parameter. In any case, we use the usual notation for differential forms (see Section A.4). An exception is the symbol $F_t^*(i(X_t)\sigma_t)$, whose meaning as a differential form on M must be clarified in part b) of the proof, since with the general preconditions X_t is not a vector field on M' and therefore the symbol $i(X_t)\sigma_t$ does not give rise to an inner product in the usual sense.

Proof. a) The statement will first be proved for the special case $M = M' = N \times I$, where N is an n–manifold, I is an interval, $F_t = \psi_t$ with $\psi_t(x, s) = (x, s + t)$ for $x \in N$, and $s \in I$. A differential form σ_t on $N \times I$ of degree k, which depends on x, s and the parameter t, can be written as

$$\sigma_t = ds \wedge a\,(x,\, s,\, t)\, dx^k + b\,(x,\, s,\, t)\, dx^{k+1},$$

where the coefficients are given in the abbreviated form

$$a(x,\, s,\, t)\, dx^k := \sum_{i_1<\ldots<i_k} a_{i_1\ldots i_k}(x,\, s,\, t)\, dx_{i_1} \wedge \ldots \wedge dx_{i_k}.$$

With this, we clearly have

$$\psi_t^*\sigma_t = ds \wedge a(x,\, s+t,\, t)\, dx^k + b\,(x,\, s+t,\, t)\, dx^{k+1},$$

and so

$$\begin{aligned}\frac{d}{dt}(\psi_t^*\sigma_t) &= ds \wedge \frac{\partial a}{\partial s}(x,\, s+t,\, t)dx^k + \frac{\partial b}{\partial s}(x,\, s+t,\, t)\, dx^{k+1}\\ &\quad + ds \wedge \frac{\partial a}{\partial t}(x, s+t, t)\, dx^k + \frac{\partial b}{\partial t}(x,\, s+t,\, t)\, dx^{k+1}.\end{aligned} \tag{1}$$

We immediately deduce that

$$\psi_t^*\left(\frac{d\sigma_t}{dt}\right) = ds \wedge \frac{\partial a}{\partial t}(x,\, s+t,\, t)\, dx^k + \frac{\partial b}{\partial t}(x,\, s+t,\, t)\, dx^{k+1}. \tag{2}$$

The vector field X_t, in this special situation, can be written as $X_t = \frac{\partial}{\partial s}$. And so we then have

$$i(X_t)\sigma_t = i\left(\frac{\partial}{\partial s}\right)\sigma_t = a(x,\, s,\, t)\, dx^k$$

and

$$\begin{aligned} d\left(i(X_t)\sigma_t\right) &= d\left(\sum_{(i)} a_{(i)}(x,\, s,\, t)\, dx_{i_1} \wedge \ldots \wedge dx_{i_k}\right) \\ &= \sum_{(i)}\left(\frac{\partial a_{(i)}}{\partial s}(x,\, s,\, t)\, ds \wedge dx_{i_1} \ldots \wedge dx_{i_k}\right. \\ &\quad \left. + \sum_{j \notin \{i_1 \ldots i_k\}} \frac{\partial a_{(i)}}{\partial x_j}(x,\, s,\, t)\, dx_j \wedge dx_{i_1} \wedge \ldots \wedge dx_{i_k}\right), \end{aligned}$$

which can be written more briefly as

$$d\left(i(X_t)\sigma_t\right) = \frac{\partial a}{\partial s}(x,\, s,\, t)\, ds \wedge dx^k + d_x a\,(x,\, s,\, t)\, dx^{k+1}.$$

Pulling back by ψ_t, we get

$$\text{(3)} \qquad \psi_t^* d\left(i(X_t)\,\sigma_t\right) = \frac{\partial a}{\partial s}(x,\, s+t,\, t)\, ds \wedge dx^k + d_x a\,(x,\, s+t,\, t)\, dx^{k+1}.$$

With the same notation,

$$d\sigma_t = -ds \wedge d_x a\,(x,\, s,\, t)\, dx^{k+1} + \frac{\partial b}{\partial s}(x,\, s,\, t)\, ds \wedge dx^{k+1} + d_x b\,(x,\, s,\, t)\, dx^{k+2}$$

is then

$$i\,(X_t)\, d\sigma_t = -d_x a\,(x,\, s,\, t) dx^{k+1} + \frac{\partial b}{\partial s}(x,\, s,\, t)\, dx^{k+1}$$

and

$$\text{(4)} \qquad \psi_t^* i\,(X_t)\, d\sigma_t = -d_x a\,(x,\, s+t,\, t)\, dx^{k+1} + \frac{\partial b}{\partial s}(x,\, s+t,\, t)\, dx^{k+1}.$$

Adding (2), (3) and (4) and equating this with (1) gives the claim for this special case.

b) The general case will now be derived from the case a), via the following decomposition:

$$
\begin{aligned}
F_t = F \circ \psi_t \circ j \quad \text{with} \quad j &: M \to M \times I, \\
m &\mapsto (m,\, 0), \\
\psi_t &: M \times I \to M \times I, \\
(m,\, s) &\mapsto (m,\, s+t), \\
F &: M \times I \to M', \\
(m,\, s) &\mapsto F_s(m)
\end{aligned}
$$

given by the diagram

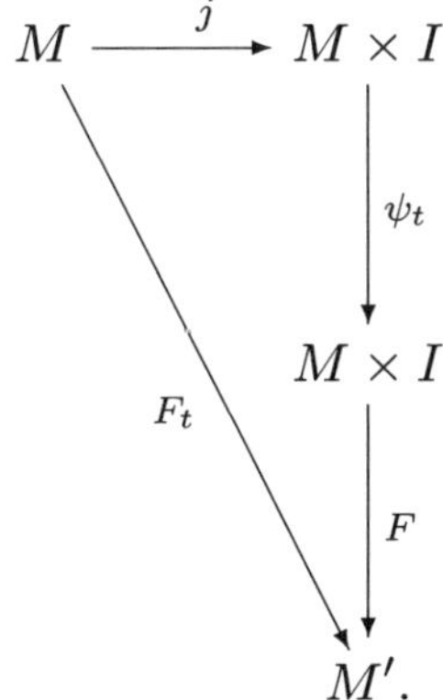

The image of the curve in $M \times I$ given by $s \mapsto (m,\, s+t)$, which for $s = 0$ passing through $(m,\, t)$ is the curve $s \mapsto F_{s+t}(m)$ passing through $F_t(m) = m'$ at $s = 0$. The tangent vector $X_t(m)$ from the tangent field along F_t at the point $m' = F_t(m) = F(m,\, t)$ is then the image of the tangent vector to the curve $\{(m,\, s+t),\, s \in I\}$, which was previously given as $\frac{\partial}{\partial s}$; that is, we have

$$
(5) \qquad (F_*)_{(m,t)} \left(\left. \frac{\partial}{\partial s} \right|_{(m,t)} \right) = X_t(m).
$$

To prepare the key conclusion, we only need to trace a vector

$$
\eta = X_m \in T_m M
$$

through the diagram which gives F_t as a composition. And so η is carried by j_* to

$$
(6) \qquad (\eta,\, 0) \in T_{(m,0)}(M \times I),
$$

by $(\psi_t)_* j_*$ to

$$
(7) \qquad (\eta,\, 0) \in T_{(m,t)}(M \times I),
$$

and finally by $(F_t)_* = F_*(\psi_t)_* j_*$ to

$$(F_t)_{*m}(\eta) \in T_{F_t(m)}M. \tag{8}$$

By Lemma 2.3, in its sharper statement that all the F_t are diffeomorphisms from M to M', X_t can be understood as a vector field on M', and for a given $(k+1)$–form σ_t on M' it gives meaning to $i(X_t)\sigma_t$ as a k–form on M'. Under the general weaker statement, $F_t^*\big(i(X_t)\big)\sigma_t$ is also meaningful and can be thought of in the following way. For $\eta_1, \ldots, \eta_k \in T_mM$ we can define

$$\begin{aligned} &F_t^*\big(i(X_t)\sigma_t\big)_m(\eta_1, \ldots, \eta_k) \\ &\qquad := (\sigma_t)_{F_t(m)}\big(X_t(m),\, (F_t)_{*m}(\eta_1), \ldots, (F_t)_{*m}(\eta_k)\big), \end{aligned}$$

and, with the presumption that $m' = F_t(m) = F(m,t)$, this with (5) and (8) gives

$$\begin{aligned} &F_t^*\big(i(X_t)\sigma_t\big)_m(\eta_1, \ldots, \eta_k) \\ &= (\sigma_t)_{F(t,m)}\left((F_*)_{(m,t)}\left(\frac{\partial}{\partial s}\Big|_{(m,t)}\right), (F_*)_{(m,t)}(\eta_1,0), \ldots, (F_*)_{(m,t)}(\eta_k,0)\right). \end{aligned}$$

Then pulling back by F to $(m,\, t)$ gives

$$\begin{aligned} F_t^*\big(i(X_t)\sigma_t\big)_m&(\eta_1, \ldots, \eta_k) \\ &= (F^*\sigma_t)_{(m,t)}\left(\frac{\partial}{\partial s}\Big|_{(m,t)}, (\eta_1,0), \ldots, (\eta_k,0)\right) \\ &= \left(i\left(\frac{\partial}{\partial s}\right)F^*\sigma_t\right)_{(m,t)}\big((\eta_1,0), \ldots, (\eta_k,0)\big)\,, \end{aligned}$$

and then pulling back by ψ_t to $(m,0)$ gives, with (7),

$$\begin{aligned} F_t^*\big(i(X_t)\sigma_t\big)_m&(\eta_1, \ldots, \eta_k) \\ &= \psi_t^*\left(i\left(\frac{\partial}{\partial s}\right)F^*\sigma_t\right)_{(m,0)}\big((\eta_1,0), \ldots, (\eta_k,0)\big), \end{aligned}$$

and, pulling back by j to m,

$$\begin{aligned} F_t^*\big(i(X_t)\sigma_t\big)_m&(\eta_1, \ldots, \eta_k) \\ &= (j^*\psi_t^*)\left(i\left(\frac{\partial}{\partial s}\right)F^*\sigma_t\right)_{(m,0)}(\eta_1, \ldots, \eta_k). \end{aligned}$$

This means that

$$F_t^* \left(i \left(X_t \right) \sigma_t \right) = j^* \psi_t^* \left(i \left(\frac{\partial}{\partial s} \right) F^* \sigma_t \right), \tag{9}$$

and so

$$F_t^* \left(i \left(X_t \right) d\sigma_t \right) = j^* \psi_t^* \left(i \left(\frac{\partial}{\partial s} \right) d \left(F^* \sigma_t \right) \right). \tag{10}$$

Now, since j^* and F^* are independent of t, we get

$$\frac{d}{dt}(F_t^* \sigma_t) = \frac{d}{dt}(j^* \psi_t^* F^* \sigma_t) = j^* \frac{d}{dt}(\psi_t^* F^* \sigma_t),$$

and applying part a) of the proof to $F^*\sigma_t$ we get, since $d\,j^* = j^* d$,

$$\begin{aligned} \frac{d}{dt}(F_t^* \sigma_t) =& j^* \psi_t^* \frac{d(F^* \sigma_t)}{dt} \\ &+ j^* \psi_t^* \left(i \left(\frac{\partial}{\partial s} \right) d \left(F^* \sigma_t \right) \right) + j^* d \left(\psi_t^* \left(i \left(\frac{\partial}{\partial s} \right) F^* \sigma_t \right) \right) \end{aligned}$$

and further, from (9) and (10),

$$\frac{d}{dt}(F_t^* \sigma_t) = j^* \psi_t^* F^* \frac{d\sigma_t}{dt} + j^* \psi_t^* F^* \left(i \left(X_t \right) d\sigma_t \right) + d \left(j^* \psi_t^* F^* \left(i \left(X_t \right) \sigma_t \right) \right).$$

With this the claim is shown. Under the stronger condition that all the F_t are diffeomorphisms, this can also be written more easily as

$$\frac{d}{dt}(F_t^* \sigma_t) = F_t^* \left(\frac{d\sigma_t}{dt} + i \left(X_t \right) d\sigma_t + d \left(i \left(X_t \right) \sigma_t \right) \right).$$

□

Remark 2.4. This formula degenerates when $M = M'$, σ_t is independent of t and F_t is the *flow* arising from a vector field X on M. Then the formula for the Lie derivative of σ along X is

$$L_X \sigma = i(X) d\sigma + d(i(X)\sigma).$$

EXERCISE 2.5. Prove the remark.

The above notion of a *flow* F_t to a given vector field X on M will be used later. Here we will only reproduce Theorem 8.1 from STERNBERG ([**St**], p. 90), which says that for every $m_0 \in M$ there exist a neighborhood U of m_0, an $\varepsilon > 0$ and a family of differentiable maps $F_t : U \to M$ with

i)

$$\begin{aligned} F: \ (-\varepsilon, \varepsilon) \times U &\rightarrow M, \\ (t, m) &\mapsto F_t(m) \text{ is differentiable,} \end{aligned}$$

ii) for $|t|$, $|s|$, $|s+t| < \varepsilon$ and $m \in U$ with $F_t(m) \in U$, we have

$$F_{s+t}(m) = F_s(F_t(m)),$$

iii) for $m \in U$, X_m is a tangent vector at $t = 0$ to the curve $t \mapsto F_t(m)$.

For the further properties we will later need we refer to Section A.4 as well as to ABRAHAM-MARSDEN ([**AM**], pp. 61–67).

The result stated at the beginning of the section is then a direct consequence of the general theorem:

THEOREM 2.6. (Darboux's Theorem) *Let ω_0 and ω_1 be two nondegenerate and closed forms of degree 2 on a 2n–dimensional manifold M with $\omega_0|_m = \omega_1|_m$ for some $m \in M$. Then there exist a neighborhood U of m and a diffeomorphism $F: U \rightarrow F(U) \subset M$ with $F(m) = m$ and $F^*\omega_1 = \omega_0$.*

Proof. The idea of the proof is to use a deformation argument to get a neighborhood U of m and a family $(F_t)_{t\in I}, I = [0,1]$, of diffeomorphisms from U to $F_t(U)$ such that

$$\begin{aligned} F_0 &= id, \\ F_1 &= F, \\ F_t(m) &= m, \quad \text{and} \\ F_t^*\omega_t &= \omega_0, \quad \text{for all } \ t \in I \quad \text{with } \omega_t := (1-t)\omega_0 + t\omega_1, \end{aligned}$$

and so, in particular, $F^*\omega_1 = \omega_0$. The F_t are realized as flows to the time dependent vector field Y_t on $F_t(U)$ with

$$(\circ) \qquad \frac{dF_t}{dt}(m') = Y_t(m') \quad \text{for all } m' \in U.$$

The determination of Y_t and thus of F_t will proceed in several steps.

a) From the equation $\omega_0 = F_t^*\omega_t$ it follows from Lemma 2.3 as a necessary condition that

$$0 = \frac{d}{dt}(F_t^*\omega_t) = F_t^*\left(\frac{d}{dt}\omega_t + i(Y_t)d\omega_t + d(i(Y_t)\omega_t)\right).$$

Since ω_0 and ω_1 are closed, so are all ω_t; that is, $d\omega_t = 0$; the necessary condition is equivalent to

$$(*) \qquad d(i(Y_t)\omega_t) = -\frac{d}{dt}\omega_t = \omega_0 - \omega_1 =: \sigma.$$

This will now be taken as the defining equation for Y_t.

b) Since σ is closed, there are, by Poincaré's lemma, a neighborhood U_1 of m and a 1–form α on U with $d\alpha = \sigma$ and $\alpha(m) = 0$. The defining equation $(*)$ becomes

$$(**) \qquad i(Y_t)\omega_t = \alpha.$$

c) We have, for all $t \in I$, that $\omega_t(m) = \omega_0(m)$, and therefore that ω_t is nondegenerate at m. It follows that this also holds in a neighborhood U_0 of m, contained in U_1. Therefore, there is in U_0 a vector field Y_t which satisfies $(**)$ and therefore also $(*)$. Because of the normalization $\alpha(m) = 0$, $Y_t(m) = 0$, we get, from the existence and uniqueness theorem for systems of ordinary differential equations, that the family $(Y_t)_{t\in I}$ of vector fields can be integrated to obtain a family $(F_t)_{t\in I}$ of diffeomorphisms which on an open neighborhood U of m satisfy $F_t(U) \subset U_0$ for all t, and satisfy $(\circ)$ with the normalizations $F_0 = id$ and $F_t(m) = m$. By construction, it follows that for $t \in I$,

$$\frac{d}{dt}(F_t^*\omega_t) = 0,$$

and then also the expected equality

$$F_t^*\omega_t = F_0^*\omega_0 = \omega_0.$$

d) The 1–form α from part b), whose existence was guaranteed by Poincaré's lemma, can be made more explicit with the application of a little analysis. Namely, this says that there is a neighborhood U_1 of m which has a *smooth retraction* $(\varphi_t)_{t\in I}$ from U_1 to $\{m\}$, that is, a family of maps

$$\varphi_t : U_1 \longrightarrow U_1 \text{ with } \varphi_1 = id,\ \varphi_t(m) = m \text{ for all } t \text{ and } \varphi_0 : U_1 \longrightarrow \{m\}.$$

For each $(\varphi_t)_{t\in I}$ we have a tangent field along φ_t in the sense of Lemma 2.3, and with its help we have

$$\sigma - \varphi_0^*\sigma = \int_0^1 \frac{d}{dt}(\varphi_t^*\sigma)dt$$

and

$$= \int_0^1 \Big(\varphi_t^*\big(\frac{d\sigma}{dt} + i(X_t)d\sigma\big) + d(\varphi_t^* i(X_t)\sigma)\Big)dt.$$

Then, since $\sigma(m) = 0, \sigma$ is closed and independent of t. We have

$$\sigma = d\alpha \text{ with } \alpha = \int_0^1 \varphi_t^*(i(X_t)\sigma)dt.$$

Then by substitution into the given formulas we get from this α the desired family $(Y_t)_{t\in I}$, respectively $(F_t)_{t\in I}$, as in c). □

In GUILLEMIN-STERNBERG ([**GS**], p. 156) there is a further *equivariant* sharpening of this theorem, which we will now describe, although this deals with the situation of manifolds with group operations, which we will study in detail later. We will consider the operation of a compact group G on the symplectic manifold M. Let $m \in M$ be a fixed point under this group operation. Then ω_0, as well as ω_1, will be G–invariant symplectic forms on M. Then there exist a G–invariant neighborhood U of m and a G–equivariant diffeomorphism F from U into M with $F(m) = m$ and $F^*\omega_1 = \omega_0$. The proof of this statement requires adding just a little bit more additional work to the proof just given.

The result stated at the beginning of this section is then a consequence of Darboux's theorem:

COROLLARY 2.7. *For each point m on the symplectic manifold (M, ω) there exist an open neighborhood U of m and a symplectomorphism F of U onto a subset $F(U)$ of $\mathbb{R}^{2n}$ equipped with the standard symplectic form ω_0.*

Proof. Here we will require a little more external yet routine analysis. This will assure the plausible existence of a diffeomorphism $F_1 : U_1 \to U$ of a neighborhood U_1 of the origin of the tangent spaces $T_mM \simeq \mathbb{R}^{2n}$ to a neighborhood U of m in M. Then $\omega_1 := F_1^*\omega$ is a symplectic form on U_1. Therefore, after a linear transformation, it can be assumed that $\omega_1|_0 = \omega_0|_0$. Darboux's Theorem 2.6 now guarantees that there exist a neighborhood $U_0 \subset U_1$ of 0 and a diffeomorphism

$$F_0 : U_0 \to F_0(U_0) \subset U_1 \text{ with } F_0(0) = 0 \text{ and } F_0^*\omega_1 = \omega_0.$$

$F := (F_1 \circ F_0)^{-1}$, which is clearly the desired symplectomorphism. □

As we have already mentioned, the coordinates given by the corollary will be called *symplectic* and will be written as (q, p).

The fact, just proved, that all symplectic manifolds of the same dimension are locally the same immediately raises questions about finding global distinguishing features. We will take a glance at some results on these questions at the end of this chapter in Section 2.7, and explore a little of what is today a very active area of research. But first we need to give a few examples of symplectic manifolds.

2.3. The cotangent bundle

The most important example of a symplectic manifold for physical applications is the cotangent bundle, $M = T^*Q$, to an n–dimensional manifold Q (see Section A.3). Q here plays the role of *configuration space* and M that of *phase space* (see Section 0.2). Such an M is necessarily a $2n$–dimensional manifold. As coordinates of a neighborhood U of a point $m \in M$ we will take, as usual, $(q,p) = (q_1, \dots, q_n, p_1, \dots, p_n)$ (which we will sometimes give in the *natural* order p, q; here we just want to make sure that it is understood that the q parametrize the configuration space Q, and that it is customary to indicate the coordinates of the base space first when speaking of a bundle). A 1–form ϑ is defined on $M = T^*Q$ and is given on U by

$$\vartheta = pdq = \sum_{i=1}^{n} p_i dq_i.$$

This form is called the *Liouville–form.* The form can also be understood, using the notation from Appendix A, as follows: ϑ is defined as a 1–form on M, given by

$$\vartheta_m(\eta) := \mu_q((\pi_*)_m \eta)$$

for $m \in M$. (Thus $m = (q, \mu_q)$, where μ is a 1–form on Q, $\eta \in T_m M$, π is the canonical projection $M = T^*Q \to Q$ carrying $m = (q, \mu_q)$ to q, and $(\pi_*)_m$ is the induced map $T_m M \to T_q Q$.) For the negative of the inner derivative $\omega := -d\vartheta$, we have, in terms of the (q,p)–coordinates,

$$\omega = -d\vartheta = dq \wedge dp = \sum_{i=1}^{n} dq_i \wedge dp_i.$$

This form is clearly closed and non–degenerate, and so defines a symplectic structure on $M = T^*Q$. Each diffeomorphism $F : Q \to Q$ naturally extends to $\hat{F} := (F^{-1})^* = F^{*-1}$ a diffeomorphism of $M = T^*Q$ to itself, which is a symplectic morphism.

EXERCISE 2.8. Prove this last comment.

2.4. Kähler manifolds

A *Kähler manifold* is, roughly speaking, a complex n–manifold (thus the transformation functions between the charts are holomorphic), equipped with a *Kähler metric*, that is, a hermitian metric for which the *associated 2–form* ω is closed. This metric was introduced by KÄHLER in 1932 [**K**], and taken up by WEIL [**We**] among others, and has, because of the peculiar properties of Kähler manifolds, become particularly significant. These manifolds form an important class of examples of symplectic manifolds. In order to introduce them, we need the material of Section 1.4. There, among other things, we defined:

- A *complex structure* on a $2n$–dimension $\mathbb{R}$ vector space V is a $J \in \operatorname{Aut} V$ with $J^2 = -\mathrm{id}_V$.
- A symplectic $\mathbb{R}$ vector space (V, ω) is called *Kähler*, if it has an ω–*compatible* complex structure J (with $J \in Sp(V)$) which satisfies $\omega\,(v,\, Jv) > 0$.

Furthermore, it follows from the discussion in [**K**] and [**We**], that for a $\mathbb{C}$ vector space W the following sets are canonically isomorphic:

- The set of *hermitian forms* h on W; that is, the set of $h : W \times W \to \mathbb{C}$ which are sesquilinear in $h(x,\, y) = \overline{h(y,\, x)}$ for all x and $y \in W$.
- The set of *symmetric* $\mathbb{R}$-bilinear forms g on W that are invariant under multiplication with i; that is, those $\mathbb{R}$-bilinear maps $g : W \times W \to \mathbb{R}$ satisfying $g(x,\, y) = g(y,\, x) = g(ix,\, iy)$ for all x and $y \in W$.
- The set of *antisymmetric* $\mathbb{R}$-bilinear forms ω on W that are invariant under multiplication by i; that is, those $\mathbb{R}$-bilinear maps $g : W \times W \to \mathbb{R}$ satisfying $\omega(x,\, y) = -\omega(y,\, x) = \omega(ix,\, iy)$ for all x and $y \in W$.

The isomorphisms are given by

$$g = \operatorname{Re} h, \quad \omega = -\operatorname{Im} h,$$

$$h(x,\, y) = g(x,\, y) + ig(x,\, iy) = \omega(ix,\, y) + i\omega(x,\, y).$$

Thus h is positive definite precisely when g is.

It is now natural to generalize the question asked in Section 1.4 of whether and in how many ways a given $\mathbb{R}$ vector space can be supplied with a complex structure for a given real manifold M. The answer to this question depends on the answer to the question of whether every real tangent space $T_mM \simeq \mathbb{R}^{2n}$ can be supplied with a complex structure J_m so that these structures vary smoothly from point to point (more precisely: they satisfy an integrability condition). We will not pursue this question further here, but will assume that M comes as a complex n–manifold. Then the tangent space $T_mM \simeq \mathbb{C}^n$, as an $\mathbb{R}$ vector space, has in a natural sense a complex structure J_m. This will correspond to the choice of local coordinates $z_j = x_j + iy_j$ $(j = 1, \ldots, n)$ and the concurrent identification of the basis

$$\frac{\partial}{\partial z_j}\Big|_m = \frac{1}{2}\left(\frac{\partial}{\partial x_j}\Big|_m - i\frac{\partial}{\partial y_j}\Big|_m\right), \qquad j = 1, \ldots, n,$$

of T_mM as $\mathbb{C}$ vector space with the basis

$$\frac{\partial}{\partial x_j}\Big|_m, \ \frac{\partial}{\partial y_j}\Big|_m, \qquad j = 1, \dots, n,$$

as $\mathbb{R}$ vector space. The multiplication by $i = \sqrt{-1}$ in T_mM as $\mathbb{C}$ vector space will be given by the map J_m,

$$\begin{aligned} J_m\left(\tfrac{\partial}{\partial x_j}\big|_m\right) &= -\ \tfrac{\partial}{\partial y_j}\big|_m, \\ J_m\left(\tfrac{\partial}{\partial y_j}\big|_m\right) &= \quad \tfrac{\partial}{\partial x_j}\big|_m, \quad j = 1, \dots, n. \end{aligned}$$

The compatibility of these structures J_m with the holomorphic coordinate transformation functions allows one to give a formulation of the above–mentioned integrability condition. We will not go into this here (see CHERN [**Ch**], p. 14, for details). We continue with the definitions.

DEFINITION 2.9. A complex n–manifold M with a symplectic structure (as real $2n$–manifold) is called a *Kähler manifold*, if for every point $m \in M$ the $\mathbb{R}$ vector space (T_mM, ω_m, J_m) is Kähler.

This is equivalent to the description given at the beginning of this section:

DEFINITION 2.10. Let M be a complex n–manifold with a *hermitian* metric g. Then M is a *Kähler manifold* if the skew symmetric bilinear form $\omega(\cdot\,,\,\cdot) := g(J\cdot, \,\cdot)$ is as an exterior 2–form, a closed differential form on M.

To this we add a few points of clarification. By a *hermitian* metric g we mean, as in AEBISCHER *et al.* ([**Ae**], p. 24), a Riemanian metric such that for every point $m \in M$, g_m is a J_m–invariant inner product on the $2n$–dimensional $\mathbb{R}$ vector space T_mM, and then J_m is compatible with g_m in the sense that

$$g_m(J_m v, \, J_m w) = g_m(v, \, w) \quad \text{for all } v, \, w \in T_mM.$$

As already seen in Section 1.4, g_m is, in the usual sense, the real part of a *hermitian scalar product* on T_mM as n–dimensional $\mathbb{C}$ vector space, namely

$$\langle\cdot\,,\,\cdot\rangle_h = h(\cdot\,,\,\cdot) = g_m(\cdot\,,\,\cdot) + i g_m(J\cdot, \,\cdot).$$

In the case when the differential form ω associated to this g_m is closed, we call it a *Kähler metric*. The findings from Section 1.4 on the connection between skew–symmetric bilinear forms ω and symmetric ones g apply here immediately to verify the equivalence of Definitions 2.9 and 2.10.

In practice we find ourselves with the following procedure: we are given a complex n–manifold M on which there is a Riemannian metric g, so that g_m and J_m are compatible at every point $m \in M$. This means a non–degenerate 2–form ω is also given. To see M as Kähler, and so also as symplectic, we must prove that $d\omega = 0$. A useful criterion in this situation is due to MUMFORD ([**Mu**], p. 87). Let G be a group of diffeomorphisms acting on M, which under the operation

$$\begin{aligned} G \times M &\longrightarrow M, \\ (g, m) &\longmapsto \phi_g(m) =: gm, \end{aligned}$$

leave the complex structure and the metric h unchanged. For $m \in M$ denote by

$$G_m = \{g \in G;\ gm = m\}$$

the isotropy group of m. Then ϕ_g induces, for each $g \in G_m$, a map

$$((\phi_g)_*)_m : T_m M \to T_m M,$$

and so a representation ϱ_m of G_m in $T_m M$, thus a homomorphism

$$\varrho_m : G_m \to \mathrm{Aut}_{\mathbb{C}}(T_m M).$$

We have

THEOREM 2.11. (Mumford's criterion) *If $J_m \in \varrho_m(G_m)$ for all $m \in M$, then $d\omega = 0$.*

Proof. Since G leaves the complex structure and the metric fixed, G also leaves ω and hence $d\omega$ unchanged. Therefore, for all $g \in G_m$ and $u, v, w \in T_m M$,

$$d\omega_m\big(\varrho_m(g)u,\ \varrho_m(g)v,\ \varrho_m(g)w\big) = d\omega_m(u, v, w).$$

Here setting $\varrho_m(g) = J_m$ and applying the formula twice yields

$$\begin{aligned} d\omega_m(u, v, w) &= d\omega_m(J_m u, J_m v, J_m w) = d\omega_m(J_m^2 u, J_m^2 v, J_m^2 w) \\ &= d\omega_m(-u, -v, -w) = -d\omega_m(u, v, w) = 0. \end{aligned}$$

□

This criterion can be used to show that complex projective space $\mathbb{P}(\mathbb{C}^{n+1}) = \mathbb{C}P^n$ is a Kähler manifold (see Section 2.6).

AEBISCHER *et al.* ([**Ae**], pp. 27 ff.) go on to prove the following *criterion.* The condition $d\omega = 0$ is equivalent to

$$\nabla_X J = 0 \quad \text{for all } X \in \Gamma(TM).$$

By ∇, we here mean the connection corresponding to the Riemannian metric g (see Section A.4); by X a vector field, thus a global section of the tangent bundle; and the complex structure J appears as a tensor field of type (1,1), thus as a global section of $T^{(1,1)}M$.

They ([**Ae**], pp. 28 ff.) then use this criterion to show that the *unit ball* in $\mathbb{C}^n$,

$$\mathbf{B}^n := \{z \in \mathbb{C}^n, ||z|| < 1\},$$

with the aid of the *Bergmann–metric* with the kernel

$$K_n(z, \varphi) := \frac{n!}{\pi^n} \frac{1}{(1 - z\overline{\varphi})^{n+1}}, \; z, \varphi \in \mathbf{B}^n,$$

is equipped with a Kähler metric.

Towards a formalism of complex differential forms: the Kähler form. In texts in which complex manifolds are the central theme (see CHERN ([**Ch**], p. 53), WEIL ([**We**], p. 41) or KÄHLER [**K**]), the description of Kähler manifolds uses the standard formalism of the real differential forms as described in Section A.4,

$$\alpha = \sum b_{i_1 \dots i_q} dx_{i_1} \wedge \dots \wedge dx_{i_q}$$

extended to the complexes. Following the Wirtinger calculus of function theory (see FISCHER–LIEB [**FL**], pp. 22–23), we assign to the complex coordinates $z_j = x_j + iy_j$ $(j = 1, \dots, n)$ the symbols

$$(*) \qquad dz_j = dx_j + idy_j, \quad d\overline{z}_j = dx_j - idy_j,$$

as well as the differential operators known from the Cauchy–Riemann differential equations,

$$\frac{\partial}{\partial z_j} = \frac{1}{2}\left(\frac{\partial}{\partial x_j} - i\frac{\partial}{\partial y_j}\right), \quad \frac{\partial}{\partial \overline{z}_j} = \frac{1}{2}\left(\frac{\partial}{\partial x_j} + i\frac{\partial}{\partial y_j}\right)$$

and define differential forms, for example

$$\Omega = \sum_{j,k} c_{jk} dz_j \wedge d\overline{z}_k \quad (c_{jk} \text{ are } \mathbb{C}\text{–valued functions.}).$$

This is called a *form of type* (1,1), since it is homogeneous of degree 1 in the dz_j and the $d\overline{z}_k$. Such a form is called *closed* when $d\Omega = 0$, where here

$$d = \partial + \overline{\partial} \quad \text{with} \quad \partial\Omega = \sum_{j,k,l} \frac{\partial c_{jk}}{\partial z_l} dz_l \wedge dz_j \wedge d\overline{z}_k$$

$$\text{and} \quad \overline{\partial}\Omega = \sum_{j,k,l} \frac{\partial c_{jk}}{\partial \overline{z}_l} d\overline{z}_l \wedge dz_j \wedge d\overline{z}_k.$$

This will now be used in an application. A complex differential manifold M of complex dimension n is called *hermitian* if TM has a hermitian structure; that is, for every point $m \in M$ the n–dimensional $\mathbb{C}$ vector space T_mM has a hermitian scalar product $\langle \; , \; \rangle_m$ assigned in a smooth way. For a chart

(φ, U) with coordinates $z = (z_1, \ldots, z_n)$ a positive definite hermitian matrix is defined, for each $m \in U$, by

$$H = (H_{jk}) \text{ with } H_{jk} = \left\langle \frac{\partial}{\partial z_j}, \frac{\partial}{\partial z_k} \right\rangle \quad (j, k = 1, \ldots, n).$$

To this matrix can then be associated the $(1,1)$–form

$$\Omega = (i/2) \sum_{j,k} H_{jk} dz_j \wedge d\overline{z}_k,$$

which CHERN called the *Kähler form.* Ω is clearly real in the sense that we have

$$\overline{\Omega} = -(i/2) \sum_{j,k} \overline{H}_{jk} d\overline{z}_j \wedge dz_k = (i/2) \sum_{j,k} H_{kj} dz_k \wedge d\overline{z}_j = \Omega.$$

As an easy variant (see below) of Definition 2.10, we have

DEFINITION 2.12. M is called a *Kähler manifold* if M is hermitian and the corresponding Kähler form Ω is closed.

Of particular importance, especially for applications of Kähler forms, is the fact that Ω is closed exactly when Ω locally has a *potential*; more precisely,

THEOREM 2.13. *Let M be hermitian with a Kähler form Ω. Then M is Kähler precisely when there is locally an $\mathbb{R}$–valued differentiable function f satisfying*

$$\Omega = i\partial\overline{\partial} f.$$

EXERCISE 2.14. Prove this. (A proof can be found in KÄHLER [**K**], as well as in CHERN ([**Ch**], p. 56).

The connection of Definition 2.10 with the real theory described above is that the Kähler form via $(*)$ can be transformed into a real differential form.

EXERCISE 2.15. By a short calculation, show that

$$\Omega = -\sum_{j<k} \beta_{jk}(dx_j \wedge dx_k + dy_j \wedge dy_k) + \sum_{j,k} \alpha_{jk} dx_j \wedge dy_k.$$

Here the $(\alpha_{jk}) = A = \operatorname{Re} H$ form a symmetric matrix, while the $(\beta_{jk}) = B = \operatorname{Im} H$ form a skew–symmetric one. We have

$$\begin{aligned} H(z, z') &= {}^t z H \overline{z}' \\ &= {}^t x A x' + {}^t y A y' + {}^t x B y' - {}^t y B x' \\ &\quad + i({}^t x B x' + {}^t y B y' + {}^t y A x' - {}^t x A y') \end{aligned}$$

This can then be likened to the earlier real theory. There we spoke of a real compatible metric g on M as a $2n$–dimensional real manifold with a complex structure J, thus with

$$g(v,w) = g(w,v) = g(Jv,Jw) \text{ for all } v,w \in T_mM \simeq \mathbb{R}^{2n}.$$

When the matrix of g is replaced by

$$g = \begin{pmatrix} A & B \\ C & D \end{pmatrix} \quad \text{and} \quad J = \begin{pmatrix} 0 & 1 \\ -1 & 0 \end{pmatrix},$$

we arrive at the condition

$$g = \begin{pmatrix} A & B \\ -B & A \end{pmatrix} \text{ with } A = {}^tA \text{ and } B = -{}^tB.$$

Then, from ${}^tv = ({}^tx, {}^ty), {}^tv' = ({}^tx', {}^ty'), {}^tz = {}^tx + i{}^ty$ and ${}^tz' = {}^tx' + i{}^ty'$, we have

$$H(z,z') = g(v,w) + ig(Jv,w).$$

The written formalism of the symmetric form g gives the assignment of the skew symmetric form as

$$\omega(v,\, w) = g(Jv,\, w).$$

After this is multiplied by $1/2$ it has the associated differential form

$$\sum_{j<k} \beta_{jk}(dx_j \wedge dx_k + dy_j \wedge dy_k) - \sum_{j,k} \alpha_{jk} dx_j \wedge dy_k.$$

This is not exactly the above Kähler form Ω for H, but rather that for $\overline{H}$. WEIL, by the way, associated to H the $(1,1)$ form

$$i/2 \sum_{j,k} h_{jk} d\overline{z}_j \wedge dz_k,$$

(WEIL ([**We**], pp. 15 and 41)), and this is then in full agreement with the real formalism.

2.5. Coadjoint orbits

This section will need to lean more heavily on the representation theory of Lie groups and algebras (see Appendix D) and on the theory of systems of differential equations on manifolds than either the preceding or the following sections will. However, courage will soon be rewarded, for we will gain a better description and the means for constructing many more manifolds; this will be needed in the discussion of the moment map and of quantization.

Coadjoint orbits arise in a natural way for any given Lie group G. The group operates via a *coadjoint representation* Ad^* on the dual space $\mathfrak{g}^*$ of the Lie algebra $\mathfrak{g}$ of G. The orbits of this action are called *coadjoint orbits* and can (under known conditions) be made into symplectic manifolds. More precisely, the goal of this section will be to discuss the following *theorem of Kostant and Souriau*:

Let G be a Lie group with $H^1(\mathfrak{g}) = H^2(\mathfrak{g}) = \{0\}$ for $\mathfrak{g} = \text{Lie } G$. (Here by $H^k(\mathfrak{g})$ we mean the k–th cohomology group of $\mathfrak{g}$ (see Appendix C)). Then there is (up to covering) a one–to–one correspondence between the symplectic manifolds with transitive G–operation and G–orbits in $\mathfrak{g}^*$.

The study of the coadjoint orbits was introduced by KIRILLOV, and the reader may find in [**Ki**], pp. 226 ff., a readable introduction to this topic. Here though, we will follow the treatment of AEBISCHER *et al.* ([**Ae**], pp. 32–39), where one may find a summary of the detailed treatment of GUILLEMIN-STERNBERG ([**GS**], pp. 172 ff.).

The coadjoint representation on $\Omega^q_l(G) \simeq \Lambda^q\mathfrak{g}^*$. For what follows, we fix G to be a Lie group and $\mathfrak{g}$ its Lie algebra, which we identify with T_eG or with the space $V_l(M)$ of all left–invariant vector fields X on G (for preliminaries, see Appendix B). Analogously, $\mathfrak{g}^*$ with T^*_eG is identified with the space of left–invariant differential forms of degree 1, and from this it then follows that the space $\Omega^q_l(G)$ of left–invariant q–forms can be identified with $\Lambda^q\mathfrak{g}^*$, the space of alternating q–forms on $\mathfrak{g}$.

Here is a good time to make the following fundamental observations.

Remark 2.16. Via this identification of $\Omega^q_l(G)$ with $\Lambda^q\mathfrak{g}^*$, the operators of exterior differentiation on $\Omega^q_l(G)$ are exactly the coboundary operators δ on $\Lambda^q\mathfrak{g}^*$ defined in Section C.2. Thus, in particular, the space $Z^2(\mathfrak{g})$ of 2–cycles in $\Lambda^2\mathfrak{g}^*$ can be identified with the space of left–invariant closed differentials on G.

Proof. i) Here, as in much of what follows, the Lie derivative acts on differential forms and vector fields of a manifold M. This is described in Sections A.4 and B.2. These actions are in fact linked according to the following formula. For $\alpha \in \Omega^q(M)$ and $X_1, \ldots, X_q \in V_l(M)$ we have (see ABRAHAM–MARSDEN ([**AM**], p. 117))

$$(*) \qquad \begin{aligned}(L_X\alpha)(X_1,\ldots,X_q) = \; & L_X(\alpha(X_1,\ldots,X_q)) \\ & -\sum_{i=1}^{q} \alpha(X_1,\ldots,[X, X_i],\ldots, X_q).\end{aligned}$$

As described in Section A.4 and as a special case of Lemma 2.3, the Lie derivative is related to inner multiplication through the formula

$$(**) \qquad L_X\alpha = d(i(X)\alpha) + i(X)d\alpha.$$

ii) Now let α be an element of $\Omega_l^1(G)$ and X, Y elements of $V_l(M)$. Then $\alpha(Y)$ is constant on G and, therefore,

$$L_X(\alpha(Y)) = 0.$$

The formula $(*)$ for $q = 1$ then says that

$$(L_X\alpha)(Y) + \alpha([X, Y]) = 0.$$

Since $i(X)\alpha = \alpha(X)$ is also constant on G, it further follows from $(**)$ that

$$(i(X)d\alpha)(Y) = -\alpha([X, Y]).$$

This can then be formulated as

$$d\alpha(X, Y) = -\alpha([X, Y]),$$

and with this we have the formula for δ, when α is taken as an element of $\mathfrak{g}$.

iii) Let ω be an element of $\Omega_l^2(G)$ and X, Y, Z elements of $V_l(G)$. Then $\omega(Y, Z)$ is constant, and therefore

$$L_X\omega(Y, Z) = 0.$$

The equation $(*)$ for $q = 2$ then gives

$$(L_X\omega)(Y, Z) + \omega([X, Y], Z) + \omega(Y, [X, Z]) = 0.$$

From $(**)$ we arrive at

$$(L_X\omega)(Y, Z) = (i(X)d\omega)(Y, Z) + d(i(X)\omega)(Y, Z),$$

which then, using the result of part ii) for $\alpha = i(X)\omega$, gives

$$(L_X\omega)(Y, Z) = d\omega(X, Y, Z) - \omega(X, [Y, Z]).$$

Putting these two statements together, we get

$$d\omega(X, Y, Z) = -\omega([X, Y], Z) + \omega([X, Z], Y) - \omega([Y, Z], X),$$

thus the formula for δ for $q = 2$.

iv) Through further and analogous iteration we get the formula for the general case. □

EXERCISE 2.17. Verify the formula $(*)$ for $q = 1$ and 2, and fill in the details for step iv) of the proof.

The coadjoint representation on $\Omega_l^q(G)$ can now be easily described: conjugation with $g_0 \in G$ acts as an inner automorphism of G:

$$\begin{aligned} \kappa_{g_0} : \quad G &\rightarrow G, \\ g &\mapsto g_0 g g_0^{-1}. \end{aligned}$$

κ_{g_0} induces a map of the tangent spaces; in particular, for the case $T_eG = \mathfrak{g}$

$$\mathrm{Ad}(g_0) := (\kappa_{g_0})_{*e} : \mathfrak{g} \rightarrow \mathfrak{g}.$$

This map gives, through a roughly similar mechanism, the homomorphism

$$\begin{array}{rlcl} \mathrm{Ad}: & G & \to & \mathrm{Aut}\ \mathfrak{g}, \\ & g_0 & \mapsto & \mathrm{Ad}\ g_0, \end{array}$$

which is called the *adjoint representation* of G. This gives rise (see Section D.1) to the coadjoint representation

$$\mathrm{Ad}^* : G \to \mathrm{Aut}\ \mathfrak{g}^*.$$

For $q > 1$, this induces a representation

$$\mathrm{Ad}^* : G \to \mathrm{Aut}\ \Lambda^q \mathfrak{g}^*,$$

which is also often denoted by the symbol $\mathrm{Ad}^{\#}$. In those cases where $\Lambda^q \mathfrak{g}^*$ is identified with the space of left invariant q–forms ω on G, we get the formula

$$\mathrm{Ad}^*(g_0)\,\omega = (\kappa_{g_0})^* \omega = (\varrho_{g_0})^* \omega$$

with ϱ the right translation on G.

Unlike Kirillov [**Ki**], we do not study the coadjoint orbits in $\mathfrak{g}^*$ directly, that is, the G–orbits in the coadjoint representation in $\mathfrak{g}^*$

$$G^{\#}\vartheta := \{\mathrm{Ad}^*(g_0)\vartheta;\ g_0 \in G\} \quad \text{for} \quad \vartheta \in \mathfrak{g}^*,$$

but rather study the orbits in $\Lambda^2 \mathfrak{g}^*$.

We begin with some notation. Let $(M,\, \omega)$ be a symplectic manifold on which G operates on the left as differentiable maps

$$G \times M \to M, \quad (g,\, m) \mapsto gm.$$

Then we may consider two maps

$$\phi_g : M \to M \quad \text{with } \phi_g(m) := gm \quad \text{for all } g \in G$$

and

$$\psi_m : G \to M \quad \text{with } \psi_m(g) := gm \quad \text{for all } m \in M.$$

In the situation where ω is fixed by each ϕ_g, $g \in G$, we then have

$$\phi_g^* \omega = \omega \quad \text{for all } g \in G.$$

We call this given operation of G on M a *symplectic operation.*

For the right translation ϱ and the left translation λ we clearly have

$$\begin{array}{rcl} \phi_g \circ \psi_m & = & \psi_m \circ \lambda_g, \\ \psi_{gm} & = & \psi_m \circ \varrho_g. \end{array}$$

With the help of ψ_m, forms can be pulled back from M to G; in particular, the closed form ω on M induces a closed form on G. More precisely, we have the following statement:

THEOREM 2.18.

a) *A symplectic group operation* $G \times M \to M$ *defines a map*

$$\begin{aligned} \Psi: \quad M &\longrightarrow Z^2(\mathfrak{g}), \\ m &\longmapsto \psi_m^* \omega, \end{aligned}$$

with

(#) $$\Psi(gm) = Ad^*(g)(\Psi(m)).$$

b) $\Psi(M)$ *is the union of* G*–orbits in* $Z^2(\mathfrak{g})$.

In the case that the operation of G is transitive, the image of $\Psi(M)$ consists of one orbit. Because of the commutation rule (#), Ψ is also called a *G–morphism.*

Proof. a) $\psi_m^* \omega$, as the image of a 2–form, is itself a 2–form on G. It is left–invariant, and so we have

$$\lambda_g^* \psi_m^* \omega = (\psi_m \circ \lambda_g)^* \omega = (\phi_g \circ \psi_m)^* \omega = \psi_m^* \circ \phi_g^* \omega = \psi_m^* \omega.$$

Ψ is a G–morphism, and so we have

$$\psi_{gm}^* \omega = (\psi_m \circ \varrho_g)^* \omega = \varrho_g^* \psi_m^* \omega = \varrho_g^* \Psi(m) = \mathrm{Ad}^*(g)\Psi(m).$$

From the fact that $\Psi(m) = \psi_m^* \omega$ is closed, we have

$$d\Psi(m) = d\,(\psi_m^* \omega) = \psi_m^* d\omega = 0.$$

b) The orbit through the point $\Psi(m)$ has the form

$$G^{\#}(m) := \{\Psi(gm) = \mathrm{Ad}^*(g)\Psi\,(m);\ g \in G\} \quad (\text{also } =: G^{\#}\Psi(m)).$$

Clearly,

$$\Psi(M) = \bigcup_{m \in M} G^{\#}(m),$$

and for a transitive operation also $\Psi(M) = G^{\#}(m)$. □

Here we now immediately pose, for $\omega \in Z^2(\mathfrak{g})$,

QUESTION 2.19. *Are there, for a given* G*–orbit* $G^{\#}\omega$ *in* $Z^2(\mathfrak{g})$*, a symplectic manifold* M *and a map* $\Psi : M \to Z^2(\mathfrak{g})$ *as above with* $G^{\#}\omega = \Psi(M)$*?*

This gives us cause to ask whether M can be constructed as a homogeneous space of the form $M = G/H$, where H is a closed Lie subgoup of G. To find such an H, the difficulty will be in showing that a symplectic form $\overline{\omega}$ on G/H can be defined, so that for the given closed $\omega \in Z^2(\mathfrak{g})$ we have

$\mathrm{pr}^*\overline{\omega} = \omega$, where pr: $G \to G/H$ is the natural projection. So let $\omega \in Z^2(\mathfrak{g})$ be a closed form on G, and set

$$\mathfrak{h}_\omega := \{X \in \mathfrak{g};\ i\,(X)\,\omega = 0\}.$$

Remark 2.20. $\mathfrak{h}_\omega$ is a subalgebra; $\mathfrak{h}_\omega = \{0\}$ when ω is non–degenerate.

Proof. The last statement follows immediately from the definition of an inner product $i\,(X)\,\omega$. To prove the first statement we need to show that for $X,\ Y \in \mathfrak{h}_\omega$ we have $[X,\ Y] \in \mathfrak{h}_\omega$. After the application of the Lie derivative, this has the same appearance as the proof of the previous remark. In detail, for $Z \in \mathfrak{g}$ in terms of the formulas $(*)$ used there, we have

$$0 = L_Y(\omega(X,Z)) = (L_Y\omega)(X,Z) + \omega([Y,X],Z) + \omega(X,[Y,Z]),$$

and in terms of $(**)$

$$L_Y\omega = i(Y)d\omega + d(i(Y)\omega) = 0,$$

implying that ω is closed. Both statements together then give, for $X, Y \in \mathfrak{h}_\omega$,

$$\omega([Y,X],Z) = 0,$$

and so $[X,Y] \in \mathfrak{h}_\omega$. □

One of the central theorems of the theory of Lie groups now says that, given such an $\mathfrak{h}_\omega$, there exists, up to covering, exactly one connected subgroup $H_\omega \subset G$ with Lie $H_\omega \simeq \mathfrak{h}_\omega$. In this case, this delivers the desired H_ω.

THEOREM 2.21. *Suppose that H_ω is closed. Then there is exactly one symplectic form $\overline{\omega}$ on $M_\omega := G/H_\omega$ with $\omega = \mathrm{pr}^*\overline{\omega}$, where*

$$\mathrm{pr} : G \to G/H_\omega = M_\omega$$

is the canonical projection.

Proof. Here again we will need some external help. We begin by showing that the condition that H_ω is closed implies that $G/H_\omega = M_\omega$ is really a differentiable manifold. We note first that the orbits aH_ω (which is an abbreviation for $ai(H_\omega)$), $a \in G$, generate the points of M_ω. At the least, this gives a locally transparent description of M_ω in coordinates, in which the orbit aH_ω of $a \in G$ in G can be seen as an *integral manifold* of a *differential system* Δ (also called a *distribution*). Here we can use a few concepts and the central statement from the theory of systems of (partial) differential equations (as can be found in the text STERNBERG ([**St**], p. 130), for example):

i) A *c–dimensional differential system* Δ on an n–dimensional differentiable manifold G (which does not necessarily need to be, as in this special case, a Lie group) is, for $1 \leq c \leq n$, a map which assigns to each $a \in G$ a c–dimensional subspace $\Delta(a)$ of T_aG:

$$G \ni a \mapsto \Delta(a) \subset T_aG \quad \text{and} \quad \dim \Delta(a) = c.$$

ii) Such a system Δ is called *smooth* if for each $a \in G$, there exist (a) a neighborhood $U(a)$ of a in G, and (b) smooth vector fields $Z_1, \ldots, Z_c$ on $U(a)$, such that for every $m \in U(a)$ the vectors $(Z_1)_m, \ldots, (Z_c)_m$ form a basis of $\Delta(p)$ (this condition takes on a much more elegant form in the language of vector bundles).

iii) A differential system Δ is said to be *involutive* if and only if Δ is smooth and

$$[X, Y] \in \Delta \quad \text{for all } X, Y \in \Delta,$$

where $X \in \Delta$ means that $X_a \in \Delta(a)$ for all $a \in G$.

iv) A submanifold $N \overset{i}{\hookrightarrow} M$ of M is called an *integral manifold* of Δ, if

$$i_*(T_qN) = \Delta\big(i(q)\big) \quad \text{for all } q \in N.$$

This thus means that the tangent space T_qN of N is, at every point $q \in N$, isomorphic to the given subspace $\Delta\big(i(q)\big)$ of the tangent spaces $T_{i(q)}G$ given by the differential system Δ. (It would perhaps be better to call this a *maximal* integral manifold, since then it would make sense to consider objects N with $i_*(T_qN) \subset \Delta\big(i(q)\big)$.)

A criterion for the existence of these maximal integral manifolds is delivered by the following:

THEOREM 2.22. (Frobenius' Theorem) *Let Δ be a c–dimensional smooth differential system on a differentiable n–manifold G. Then:*

a) *Δ is involutive exactly when every point of $a \in G$ lies in an integral manifold $N = N_a$.*
b) *Should a) be satisfied, there are local coordinates $x_1, \ldots, x_n$ so that the integral manifold has the form*

$$N = \{x | x_i \textit{ is constant, } i = 1, \ldots, k\}, \quad k = n - c.$$

This form of Frobenius' theorem, so written with the help of vector fields. has a mirror image in the world of differential forms (which historically came first). A proof of Frobenius' theorem in this form can be found in KÄHLER [**K1**].

v) Now we return yet again to the special situation of the proof of Theorem 2.21. In particular, we want to interpret the Lie algebras as subspaces of tangent spaces in order to define a differential system Δ by

$$\Delta(e) := \mathfrak{h}_\omega \quad \text{for the identity element } e \in G$$

and

$$\Delta(a) := (\lambda_a)_*(\mathfrak{h}_\omega), \quad \text{for } a \in G \text{ with the left translation } \lambda_a.$$

From Remark 2.20, it immediately follows that this Δ is involutive and, from iv), that every point $a \in G$ has an integral manifold N_a. These are clearly given by $N_a = aH_\omega$, and further, from Frobenius' theorem (part b), N_a, at every $a \in G$, can be locally written as

$$N_a = \{x | x_i \text{ is constant }, i = 1, \ldots, k\}, \quad k = n - \dim \mathfrak{h}_\omega.$$

The tangent space $T_x N_a$ then has as basis

$$\frac{\partial}{\partial x_{k+1}}, \ldots, \frac{\partial}{\partial x_n}.$$

Now the given closed 2–form ω on G has the form

$$\omega = \sum_{i<j}^{n} a_{ij}(x)\, dx_i \wedge dx_j.$$

It follows from the construction of $\mathfrak{h}_\omega$ that $i(X)\omega = 0$ for $X \in \mathfrak{h}_\omega$, thus, in particular, for $X = \frac{\partial}{\partial x_i}$, $i = k+1, \ldots, n$. This and the fact that $d\omega = 0$ is closed show that ω, in these coordinates, can be dependent only on $x_1, \ldots, x_k$. Thus

$$\omega = \sum_{i<j}^{k} a_{ij}(x_1, \ldots, x_k)\, dx_i \wedge dx_j.$$

With this, we are essentially done. The integral manifolds N_a of Δ are the fibers of the projection $\mathrm{pr} : G \to M_\omega = G/H_\omega$, and M_ω is described in the local coordinates $(x_1, \ldots, x_k)$. For

$$\overline{\omega} := \sum_{i<j}^{k} a_{ij}(x_1, \ldots, x_k)\, dx_i \wedge dx_j$$

we naturally have $\mathrm{pr}^*\overline{\omega} = \omega$. It is now routine to see that in this manner also a global form $\overline{\omega}$ on M_ω with the desired properties can be given. □

In GUILLEMIN-STERNBERG ([**GS**], p. 174), Theorem 2.21 is given for a somewhat more general case. In this generality it lays the groundwork for the so-called *symplectic reduction*, which has the goal of projecting a given manifold onto a (smaller) symplectic manifold by exploiting *additional* *symmetries.* We will give a presentation of this material in Section 4.3. The proof uses the same ideas and is only a little harder.

THEOREM 2.23. *Let M be a differentiable manifold and ω a closed 2–form on M. Then the vector space of vector fields X on M with*

$$(*) \qquad i\,(X)\,\omega = 0$$

is closed with respect to the Lie bracket. In the case that the dimension of the space of those X satisfying $(*)$ is constant at every point $m \in M$, ω defines an integrable differential system Δ. Should there, further, be given a manifold M_0 and a submersion $\varrho : M \to M_0$ (that is, ϱ is a differentiable surjection which induces surjective maps on the tangent spaces) so that the integral manifolds of Δ are the fibers of ϱ, then there is exactly one symplectic form ω_0 on M_0 with $\varrho^*\omega_0 = \omega$.

The symplectic manifold of an orbit $G^{\#}\omega$. With this last theorem, we can now answer Question 2.19. For a given orbit $G^{\#}\omega$ in $Z^2(\mathfrak{g})$, H_ω and (since under the presumptions H_ω is closed) the manifold M_ω are fixed. Then $G^{\#}\omega$ is the image of M_ω under the map Ψ. Thus

$$\Psi\,(M_\omega) = G^{\#}\omega.$$

Proof. When $m_0 = eH_\omega$ is interpreted as a point of M_ω, we have, for $a \in G$, with the canonical projection $\mathrm{pr} : G \to M_\omega = G/H_\omega$ and in the notation introduced at the beginning of the section,

$$\mathrm{pr}\ a = aH_\omega = am_0 = \psi_{m_0}(a);$$

thus $\mathrm{pr} = \psi_{m_0}$. When we further define $\Psi : M_\omega \to Z^2(\mathfrak{g})$ as in Theorem 2.18 by

$$\Psi\,(m) := \psi_m^*\,\overline{\omega}$$

for ω as in Theorem 2.21, we get

$$\Psi\,(m_0) = \ \mathrm{pr}^*\overline{\omega} = \omega.$$

Since Ψ is a G–morphism, we conclude that

$$\Psi\,(am_0) = \mathrm{Ad}^*(a)\,\Psi(m_0) = \mathrm{Ad}^*(a)\,\omega$$

and so

$$\Psi\,(M_\omega) = G^{\#}\omega,$$

since G operates transitively on M_ω. □

Now that this construction of M_ω for a given G–orbit of $\omega \in Z^2(\mathfrak{g})$ for closed H_ω has positively answered Question 2.19, we naturally turn to the question of its uniqueness

QUESTION 2.24. *Is there more than one homogeneous symplectic manifold attached to a given orbit* $G^{\#}\omega$*?*

To answer this question, we let a symplectic manifold $M = G/H$ with symplectic form Ω be given, where H in G is a closed but not necessarily connected subgroup, and with a symplectic transitive operation $G \times M \to M$. Then this operation induces, for all $b \in G$, a map

$$\begin{aligned} \psi_{bH} : \quad G &\to M, \\ a &\mapsto \psi_{bH}(a) = abH. \end{aligned}$$

By Theorem 2.18, there further exists a map

$$\Psi : M \to Z^2(\mathfrak{g}).$$

We then set

$$\omega := \Psi\,(eH) = \psi_{eH}^{*}\,\Omega,$$

and because of the transitivity of the G–operation we have that

$$\Psi(M) = \{\mathrm{Ad}^*(a)\,\omega;\ a \in G\} = G^{\#}\omega$$

consists of but a single G–orbit. The uniqueness can now be positively answered, since, by the details of the proof of Theorem 2.21, we have that M_ω is the same as the given M. To see this, as in the proof of Theorem 2.21, for $\mathfrak{g} = \mathrm{Lie}\ G$ let

$$\mathfrak{h}_\omega := \{X \in \mathfrak{g};\ i\,(X)\,\omega = 0\}.$$

Thus

$$\mathfrak{h}_\omega = \{X \in \mathfrak{g};\ \psi_{eH}^{*}\,\Omega\,(X,\ \cdot\ ) = 0\},$$

and since Ω is nondegenerate, we also have

$$\mathfrak{h}_\omega = \{X \in \mathfrak{g};\ (\psi_{eH})_* X = 0\}.$$

But this shows that $\mathfrak{h} := \mathrm{Lie}\ H = \mathfrak{h}_\omega$, since

$$\begin{aligned} (\psi_{eH})_*\big|_{\mathfrak{g}} : \mathfrak{g} = T_eG &\to T_{eH}M = \mathfrak{g}/\mathfrak{h}, \\ X &\mapsto X + \mathfrak{h} \end{aligned}$$

means that $X \in \mathfrak{h}$ is equivalent to $(\psi_{eH})_* X = 0$. As things are now arranged, H_ω is the *connected* subgroup of G associated to $\mathfrak{h}_\omega = \mathfrak{h}$, thus the connected component of unity in H. And so $M_\omega = G/H_\omega$, when not the same as $M = G/H$, is then shown to be a covering of M. With this, we have now proved

THEOREM 2.25. *Homogeneous symplectic manifolds* G/H *are, up to covering, parametrized by the* G*–orbits* $G^{\#}\omega$ *in* $Z^2(\mathfrak{g})$*, if a given* ω *gives rise to a closed subgroup* H_ω *in* G*.*

From these results the statement made at the beginning of the section quickly follows:

THEOREM 2.26. (Kostant–Souriau) *For* $H^1(\mathfrak{g}) = H^2(\mathfrak{g}) = \{0\}$ *there is, up to covering, a one–to–one correspondence between the symplectic manifolds for* G *and the* G*–orbits in* $\mathfrak{g}^*$.

Proof. i) The cohomology groups $H^k(\mathfrak{g})$ are introduced in Section C.2. $H^2(\mathfrak{g}) = \{0\}$ says that for every $\omega \in Z^2(\mathfrak{g})$ there is a $\beta \in \mathfrak{g}^*$ with $d\beta = \omega$. In a similar vein, $H^1(\mathfrak{g}) = \{0\}$ is equivalent to $Z^1(\mathfrak{g}) = B^1(\mathfrak{g})$. But $B^1(\mathfrak{g}) = d\,(\Lambda^\circ \mathfrak{g}^*) = \{0\}$ (since $\Lambda^\circ \mathfrak{g}^* = \mathbb{R}$), and therefore $d\beta = d\beta'$ implies $\beta' = \beta$, and so $\beta \in \mathfrak{g}^*$ with $d\beta = \omega$ uniquely given by ω.

ii) This says that there is a one–to–one correspondence between the G–orbits in $Z^2(\mathfrak{g})$

$$G^{\#}\omega = \{\mathrm{Ad}^*(a)\omega;\ a \in G\}$$

and the G–orbits in $\mathfrak{g}^*$

$$G^{\#}\beta = \{\mathrm{Ad}^*(a)\beta;\ a \in G\}.$$

Thus, a bijective map

$$\mathrm{Ad}^*(a)\beta \mapsto \mathrm{Ad}^*(a)\omega \text{ for all } a \in G$$

is given, and it follows that

$$d\left(\mathrm{Ad}^*(a)\beta\right) = d\left(\varrho_a^*\beta\right) = \varrho_a^*\,\omega = \mathrm{Ad}^*(a)\,\omega.$$

iii) The isotropy group

$$G_\beta := \{a \in G;\ \mathrm{Ad}^*(a)\,\beta = \beta\}$$

is *per se* closed. Then H_ω is also shown to be closed as soon as it is shown that H_ω is the connected component of unity of G_β. We have

$$\mathfrak{g}_\beta := \mathrm{Lie}\ G_\beta = \{X \in \mathfrak{g};\ L_X\beta = 0\},$$

and thus $\beta = \mathrm{Ad}^*(a)\,\beta = \varrho_a^*\beta$ is, by the definition of the Lie derivative, for $a = \exp X$ equivalent to $L_X\beta = 0$. This is further equivalent to $i\,(X)\,\omega = 0$. Then for $X \in \mathfrak{g}$, Lemma 2.3 says that, in connection with the formula (**) in the proof of Remark 2.16,

$$L_X\beta = i(X)d\beta + d\left(i(X)\beta\right);$$

thus

$$L_X\beta = i\,(X)\,\omega,$$

since $i\,(X)\,\beta$ is locally constant (because X and β both are left–invariant). From this we now get that

$$\mathfrak{g}_\beta = \{X \in \mathfrak{g};\ i\,(X)\,\omega = 0\} = \mathfrak{h}_\omega,$$

and so H_ω is contained in G_β and therefore closed. $\square$

A further criterion that $M_\omega = G/H_\omega$ really be a manifold was discovered by CHU. It says that H_ω is closed when G is simply connected (for a proof, see GUILLEMIN-STERNBERG ([**GS**], p. 179)). Of greater practical significance for giving explicit symplectic forms is the following theorem, where a coadjoint orbit in $\mathfrak{g}^*$ also shows up.

THEOREM 2.27. *Let G be a Lie group with $H^1(\mathfrak{g}) = H^2(\mathfrak{g}) = \{0\}$, $\beta \in \mathfrak{g}^*$, $\omega = d\beta$, and*

$$M_\beta = G/H_\omega \simeq G^\# \omega \simeq G^\# \beta$$

the associated symplectic manifold to the form $\overline{\omega}$. Then it follows that

$$\overline{\omega}\,(X_\beta,\, Y_\beta) = -\beta\left([X,\, Y]\right)$$

with

$$X_\beta := \mathrm{pr}_* X, \quad Y_\beta := \mathrm{pr}_* Y \quad \textit{for } X, Y \in \mathfrak{g}.$$

Here again, pr: $G \to G/H_\omega$ *is the canonical projection.*

Proof. From Theorem 2.21, $\overline{\omega}$ as a symplectic form on G/H_ω is uniquely determined by $\mathrm{pr}^*\,\overline{\omega} = \omega$. On the one hand, we have

$$\overline{\omega}\,(X_\beta,\, Y_\beta) = \overline{\omega}\,(\mathrm{pr}_* X,\, \mathrm{pr}_* Y) = (\mathrm{pr}^*\,\overline{\omega})(X,\, Y) = \omega\,(X,\, Y).$$

On the other hand, as given in step ii) of the proof of Remark 2.16,

$$-\beta\left([X,\, Y]\right) = d\beta\,(X,\, Y) = \omega\,(X,\, Y),$$

which can here be recognized as the formula for the coboundary operator δ in the cohomology theory in Section C.2. □

We close this section with an example which should give a first feeling for the power of Theorem 2.27: one can recover the volume form ω_0 of the two–dimensional sphere S^2 in the following way.

EXERCISE 2.28. Take $G = SO(3)$ and identify S^2 with an orbit $G^\# \beta$ for $\beta \in \mathfrak{g}^* = \mathfrak{so}(3)^*$, and use Theorem 2.27 to define a symplectic form $\overline{\omega}$ on $M_\beta = SO(3)/SO(2) \simeq S^2$.

Hint. As indicated in B.2 and used more explicitly in Example 4.22, the elements $B \in \mathfrak{g} = \mathfrak{so}(3)$ can be identified with $b \in \mathbb{R}^3$ in such a way that the Lie products $[B, B']$ correspond to the vector products $b \times b'$. In this way, we come up with a form

$$\omega = x_1 dx_2 \wedge dx_3 + x_2 dx_3 \wedge dx_1 + x_3 dx_1 \wedge dx_2$$

which, if we put coordinates on S^2 by

$$\begin{aligned} x_1 &= \cos\vartheta \cos\phi, \\ x_2 &= \cos\vartheta \sin\phi, \\ x_3 &= \sin\vartheta, \end{aligned}$$

pulls back to

$$\omega_o = \cos\vartheta d\phi \wedge d\vartheta.$$

Remark 2.29. In the same way, one can construct symplectic manifolds as orbits $G^{\#}\beta$ for the Galilei and the Poincaré group which are of great importance for physics. One is tempted to do it here, but it would take too much space. The interested reader may consult GUILLEMIN-STERNBERG [**GS**], pp. 122–130, 437–445, or SOURIAU [**So**], pp. 144–192, and also the thesis of M.A. EL GRADECHI [**EG**].

2.6. Complex projective space

From the treatment of coadjoint orbits, one can build the moment map as in GUILLEMIN-STERNBERG [**GS**]. But we will deal with this central topic later. Now we turn our attention to another concrete example which will serve to make the previous concepts and techniques somewhat more explicit. Later, the study of the moment map will give us cause to consider further examples.

Let $\mathbb{P}^n = \mathbb{P}^n(\mathbb{C}) = \mathbb{P}(\mathbb{C}^{n+1})$ be complex projective space, thus the space of all (complex!) lines passing through the origin in $\mathbb{C}^{n+1}$, and denote by π the map

$$\begin{array}{rrcl} \pi : & \mathbb{C}^{n+1}\backslash\{0\} & \rightarrow & \mathbb{P}^n, \\ & z = (z_0, \ldots, z_n) & \mapsto & z_\sim. \end{array}$$

As in Section A.1, $z_\sim$ will mean the equivalence class gotten by setting $z \sim z'$ whenever $z'_i = \lambda z_i$, $i = 0, \ldots, n$, for a $\lambda \in \mathbb{C}^*$. $\mathbb{P}^n$ is now an example of a real symplectic $2n$–manifold. This can be shown in a variety of ways; we shall do so by demonstrating that $\mathbb{P}^n$ is Kähler (and therefore, from Section 2.4, symplectic). This itself can be demonstrated in a variety of ways.

i) In MUMFORD ([**Mu**], pp. 86-87) the following metric is introduced on $\mathbb{P}^n$. Let $V = V_{Herm}$ denote the space of $(n+1)$–rowed hermitian matrices,

$$V = \{A \in M_{n+1}(\mathbb{C});\ {}^tA = \overline{A}\},$$

thought of as an $\mathbb{R}$ vector space, which then has dimension $(n+1)^2$. A differentiable map ϕ can be given by

$$\begin{array}{rcl} \phi : \mathbb{P}^n & \hookrightarrow & V, \\ z_\sim & \mapsto & (A_{ij}) = (z_i\overline{z}_j) \quad \text{for } z \text{ with } |z|^2 = 1. \end{array}$$

The group $U(n+1) = \{S \in GL_{n+1}(\mathbb{C});\ {}^tS\overline{S} = E_{n+1}\}$ operates on V by conjugation and on $\mathbb{P}^n$ by

$$(S, z_\sim) \mapsto S(z_\sim) := (Sz)_\sim.$$

ϕ is then equivariant for the operation:

$$\phi\left(S\left(z_{\sim}\right)\right)=S\phi\left(z_{\sim}\right)S^{-1}.$$

V has a $U(n+1)$–invariant positive definite symmetric standard bilinear form q, given by

$$q\left(A,\,B\right)=\operatorname{tr}(AB).$$

ϕ induces at every point $z_{\sim}=\pi\left(z\right)$ a map of the tangent spaces

$$(\phi_*)_{\pi(z)}:T_{\pi(z)}\mathbb{P}^n\rightarrow T_{\phi(z_{\sim})}V\simeq V.$$

By *pulling back*, this gives a $U(n+1)$–invariant Riemannian metric on $\mathbb{P}^n$ as real manifold. A rather brutal computation (as conceded by Mumford) shows that the associated (see Section 2.4) hermitian metric on the complex manifold, called the *Fubini–Study metric*, in the coordinates

$$x_i=z_i/z_0,\quad i=1,\ldots,n,\ \text{for } z_0\neq 0,$$

has the form

$$ds^2=\frac{\sum dx_i d\overline{x}_i}{1+\sum|x_i|^2}-\frac{(\sum x_i d\overline{x}_i)(\sum\overline{x}_i dx_i)}{(1+\sum|x_i|^2)^2}.$$

Here we give a few steps of the calculation for the enlightment of the interested reader. To

$$\begin{array}{ccccc}\mathbb{C}^n & \overset{i_0}{\hookrightarrow} & \mathbb{P}^n(\mathbb{C}) & \overset{\phi}{\longrightarrow} & V,\\ x & \mapsto & \begin{pmatrix}1\\ x\end{pmatrix}_{\sim}=z_{\sim} & \longmapsto & \dfrac{1}{1+|x|^2}\begin{pmatrix}1\\ x\end{pmatrix}(1,\,{}^t\overline{x}),\end{array}$$

with x as column, is associated the map of tangent spaces

$$\mathbb{C}^n\simeq T_x\mathbb{C}^n\xrightarrow{i_{0*}}T_{i_0(x)}\mathbb{P}^n\xrightarrow{\phi_*}T_{\phi i_0(x)}V\simeq V,$$

which takes the tangent vector u to the curve $\gamma\left(t\right)=x+ut$ in $\mathbb{C}^n$ to the tangent vector to the image curve $(\phi i_0\gamma)(t)$ at $t=0$ in the space V. This vector can be transformed (with $\langle x,\,u\rangle=\Sigma x_i\overline{u}_i$) to

$$\alpha\left(u\right):=\phi_* i_{0*}u=$$

$$\frac{1}{1+|x|^2}\left(\begin{pmatrix}0\\ u\end{pmatrix}(1,\,{}^t\overline{x})+\begin{pmatrix}1\\ x\end{pmatrix}(0,\,{}^t\overline{u})-\frac{\langle u,\,x\rangle+\langle x,\,u\rangle}{1+|x|^2}\begin{pmatrix}1\\ x\end{pmatrix}(1,\,{}^t\overline{x})\right).$$

Then as a bilinear form on $T_{i_0(x)}\mathbb{P}^n$ we take the pullback of the bilinear form living on V; thus

$$g\left(u,u'\right)=\operatorname{tr}\bigl(\alpha\left(u\right)\cdot\alpha\left(u'\right)\bigr).$$

From this, after a little calculation involving substitutions of the form

$$\operatorname{tr}\left(\begin{pmatrix}1\\ x\end{pmatrix}(1,\,{}^t x)\begin{pmatrix}1\\ \overline{u}\end{pmatrix}(1,\,{}^t\overline{u})\right)=(1+\langle x,\,u\rangle)^2,$$

we arrive at

$$g(u, u') = \frac{\langle u, u'\rangle + \langle u', u\rangle}{1+|x|^2} - \frac{\langle u, x\rangle\langle x, u'\rangle + \langle x, u\rangle\langle u', x\rangle}{(1+|x|^2)^2}.$$

This can now be recognized as the two-fold real part of the hermitian metric given above (after substitution of $u = e_i$ and $u' = e_j$).

ii) In AEBISCHER *et al.* ([**Ae**], p. 40), this metric is carried over to the definition of $\mathbb{P}^n$ as the space of one–dimensional subspaces of $\mathbb{C}^{n+1}$. Then the map

$$\begin{array}{ccccc} \mathbb{C}^{n+1}\backslash\{0\} & \xrightarrow{\pi} & \mathbb{P}^n \supset U_0 = \{z_\sim,\ z_0 \neq 0\} & \xrightarrow{\varphi_0} & \mathbb{C}^n, \\ z & \longmapsto & z_\sim & \longmapsto & x, \end{array}$$

with $x_i = z_i/z_0$ for $(i = 1, \ldots, n)$, induces the map of tangent spaces

$$\begin{array}{ccccc} \mathbb{C}^{n+1} = T_z\mathbb{C}^{n+1} & \xrightarrow{\pi_*} & T_{z_\sim}\mathbb{P}^n & \xrightarrow{\varphi_{0*}} & T_x\mathbb{C}^n = \mathbb{C}^n, \\ \xi & \longmapsto & \pi_*(\xi) & \longmapsto & (\varphi_0\pi)_*(\xi) =: \zeta. \end{array}$$

Since any $\xi \neq 0$ can be interpreted as a tangent vector to the curve

$$t \mapsto \gamma(t) = z + t\xi \quad \text{in } \mathbb{C}^{n+1},$$

it can be shown that ζ is the tangent vector at $t = 0$ to the image curve

$$t \mapsto \gamma_0(t) = (\varphi_0\pi\gamma)(t) \quad \text{in } \mathbb{C}^n$$

via the transformation

$$\zeta = (\zeta_1, \ldots, \zeta_n) \quad \text{with } \zeta_i = \xi_i/z_0 - \xi_0 x_i/z_0.$$

For two such tangent vectors ζ and ζ' , we can, in the following "natural" way, find a hermitian scalar product $\langle\langle\, ,\, \rangle\rangle$. On $\mathbb{C}^{n+1}$ the usual hermitian scalar product is given by the formula

$$\langle \xi,\, \xi'\rangle = \sum_{i=0}^{n} \xi_i \overline{\xi}'_i.$$

Denote by W the orthogonal complement in $T_z(\mathbb{C}^{n+1}) \simeq \mathbb{C}^{n+1}$ of the lines passing through $z\mathbb{C}$. Then

$$(z\mathbb{C})^\perp =: W \cong \mathbb{C}^n,$$

and W can be identified with $T_x\mathbb{C}^n = \mathbb{C}^n$ since $\pi_*(\mathbb{C}^{n+1}) = \pi_*(W)$. So we have $\xi = cz + \eta$ for a uniquely defined $c = \langle \xi, z\rangle / \langle z, z\rangle \in \mathbb{C}$ and $\eta \in W$ with

$$(\varphi_0\pi)_*(\xi) = (\varphi_0\pi)_*(\eta) = \zeta.$$

Clearly,

$$\eta = \frac{\langle z,\, z\rangle\xi - \langle \xi,\, z\rangle z}{\langle z,\, z\rangle},$$

and it follows that

$$\langle \eta, \eta' \rangle = \frac{\langle \xi, \xi' \rangle \langle z, z \rangle - \langle \xi, z \rangle \langle z, \xi' \rangle}{\langle z, z \rangle}.$$

Now as hermitian scalar product for two tangent vectors in $T_{\pi(z)}\mathbb{P}^n$ with the coordinate vectors ζ, respectively ζ', we take

$$\langle\langle \zeta, \zeta' \rangle\rangle = \frac{\langle \eta, \eta' \rangle}{\langle z, z \rangle}.$$

Here we can substitute for ξ, respectively ξ', as well as z an $(n+1)$–tuple with $(\varphi_\circ\pi)_*(\xi) = \zeta$ as well as $\varphi_0\pi\,(z) = x$ (that is, $\xi_i = \zeta_i z_0,\ \xi_0 = 0$), and then we get straightaway the form of the metric given above in i).

The formalism developed in Section 2.4 can now be applied to

$$h\,(\ ,\ ) := \operatorname{Re}\langle\langle\ ,\ \rangle\rangle.$$

This then says that this is a Riemannian metric on $\mathbb{P}^n$ which is J–invariant; that is, for the complex structure J for which $J\big|_{\pi(z)}$ operates on $T_{\pi(z)}\mathbb{P}^n$ as multiplication by $i = \sqrt{-1}$ on W, we have

$$h_p(J_p v,\ J_p w) = h_p(v,\ w) \quad \text{for } p = \pi\,(z).$$

To show that the exterior 2–form ω associated to h is closed, one uses Mumford's criterion from Section 2.4. To this end, let G be the group of diffeomorphisms of $\mathbb{P}^n$ which leave the complex structure and h invariant; thus $G = SU(n+1)$. (Here, we are exploiting the classical isomorphism

$$\mathbb{P}^n \simeq SU(n+1)/U(n).)$$

It is then observed that

$$SU(n+1)_{\pi(z)} \simeq U(W)$$

and that the representation ϱ_p in Mumford's criterion, Theorem 2.11, is here given by

$$\varrho_{\pi(z)} : SU\,(n+1)_{\pi(z)} \to U\,(W).$$

Then we clearly have $J\big|_{\pi(z)} = iE_n \in U(W)$, and the criterion says that

$$\omega\,(\cdot\,,\ \cdot) = h\,(J\cdot,\ \cdot)$$

is closed, thus that $\mathbb{P}^n$ is Kähler. Mumford proves this directly in [**Mu**], p. 87. In AEBISCHER *et al.* [**Ae**] this is crystalized to the criterion of Theorem 2.11.

iii) Thus with a little calculation one not only arrives at the metric but also gets a good feeling as to its origin. On the other hand, one may arrive at the metric a little more quickly using the formalism of Definition 2.12, by giving a potential from which the Kähler form is gotten through differentiation. And this is given on $U_i = \{(z)_\sim, z_i \neq 0\} \subset \mathbb{P}^n$, with the coordinates $x \in \mathbb{C}^n$, by

$$f = \log K \quad \text{with} \quad K(x, \overline{x}) = 1 + \sum_{j=1}^{n} x_j \overline{x}_j.$$

Then, in the formalism of Section 2.4, this can be written as

$$\partial\overline{\partial} f = \sum_{j=1}^{n} \frac{dx_j \wedge d\overline{x}_j}{K} - \sum_{j,k} \frac{\overline{x}_j dx_j \wedge x_k d\overline{x}_k}{K^2},$$

and this can be interpreted as a Kähler form from a hermitian matrix, which belongs to the Fubini–Study metrics given earlier.

iv) It is then pretty nice to demonstrate (AEBISCHER *et al.* ([**Ae**], pp. 41 ff.)) that $\mathbb{P}^n$ can also be realized as a coadjoint orbit. Here again the underlying group is naturally $G = SU(n+1)$. It can be shown that

$$\mathfrak{g} = \mathfrak{su}\,(n+1) = \{A \in \mathfrak{gl}\,(n+1, \mathbb{C});\ {}^tA = -\overline{A},\ \text{tr } A = 0\}$$

is the associated Lie algebra. On the vector space of these traceless skew hermitian matrices, we can give a scalar product (,) by the formula

$$(A, B) = \text{Re}(\text{tr } A^t\overline{B}).$$

This makes possible an identification

$$\begin{aligned} \mathfrak{g} &\to \mathfrak{g}^*, \\ A &\mapsto \varphi_A \quad \text{with } \varphi_A(B) := (A, B), \end{aligned}$$

which we will consider fixed in what follows.

Now we consider the coadjoint orbit in $\mathfrak{g}^*$,

$$G^\# B = \{\text{Ad}^*(a)\, B;\ a \in G\} = \{a^{-1}Ba;\ a \in G\},$$

for the special case $B = B_0$ with

$$B_0 = iE_-, \quad E_- = \begin{pmatrix} (1/n)\, E_n & 0 \\ 0 & -1 \end{pmatrix}.$$

Then we have

$$G^\# B_0 \simeq SU(n+1)/U(n) \simeq \mathbb{P}^n,$$

and this is $a^{-1}E_-a = E_-$ precisely for

$$a = \begin{pmatrix} T & 0 \\ 0 & t \end{pmatrix}, \quad t = (\det T)^{-1} \text{ and } T \in U(n).$$

Theorem 2.27 now says that the symplectic form ω on $\mathbb{P}^n$ is given by

$$\omega\,(X, Y) := (-B, [A^X, A^Y])$$

for $A^X, A^Y \in \mathfrak{g}$ tangent vectors such that

$$\mathrm{pr}_* A^X = X, \ \mathrm{pr}_* A^Y = Y \text{ for } \mathrm{pr} : SU(n+1) \to SU(n+1)/U(n).$$

2.7. Symplectic invariants (a quick view)

On the basis of Darboux's Theorem 2.6, all symplectic manifolds of a fixed dimension are locally isomorphic. Thus to differentiate between them we must find attributes of the global object. Here we will give only a hint of two such attributes. A more precise and fuller account of this active area of research can be read in HOFER–ZEHNDER ([**HZ**], chapter 4) and in AEBISCHER *et al.* ([**Ae**], chapter 6). Both treatments rely on the earlier work of GROMOV, who during his study of symplectic embeddings [**Gr**] formulated the following statement.

THEOREM 2.30. (Gromov's squeezing theorem) *For r, $R \in \mathbb{R}_{>0}$, let*

$$\mathbf{B}(r) := \{(x, y) \in \mathbb{R}^{2n}; |x|^2 + |y|^2 < r^2\}$$

be the open ball in $\mathbb{R}^{2n}$ and $\mathbf{Z}(R) := \{(x, y) \in \mathbb{R}^{2n}; x_1^2 + y_1^2 < R^2\}$ an open cylinder. Let ϕ be a symplectic embedding of $\mathbf{B}(r)$ defined by

$$\phi(\mathbf{B}(r)) \subset \mathbf{Z}(R).$$

Then

$$R \geq r.$$

EXERCISE 2.31. Prove this for the case that ϕ is linear.

Pseudoholomorphic curves. A proof of the above theorem is given in GROMOV [**Gr**] and uses the concept of pseudoholomorphic curves. This concept, along with some important applications and a proof of Gromov's theorem, is given by AEBISCHER *et al.* ([**Ae**], chapter 6). Let M be an almost–complex manifold (that is, there is a $J \in \mathrm{Aut}(TM)$ with $J^2 = -1$ (see also Section 2.4)), and let Σ be a Riemannian surface, thus a one–dimensional complex manifold, which is an almost–complex manifold under the map of tangent spaces induced by multiplication by i on Σ.

DEFINITION 2.32. A *pseudoholomorphic curve* is a smooth map $f : \Sigma \to M$ for which the induced map

$$Tf := f_* : T\Sigma \to TM$$

is (i, J)–linear; that is, it satisfies $f_* \circ i = J \circ f_*$.

Here the pseudoholomorphic curves are particulary interesting when there is a symplectic structure ω on M which is compatible with the almost–complex structure. As an example we may pick out a statement appearing in AEBISCHER *et al.* ([**Ae**], Lemma 6.3.6 on p. 128). Let $f : \Sigma \to M$ be a pseudoholomorphic curve, μ a metric on Σ in the conformal class of the complex structure i on Σ, and g a metric on M. The smooth map $f : \Sigma \to M$ induces the map $f_* : T\Sigma \to TM$, and so, at each point, a homomorphism

$$Tf_z = f_{*z} : T_z\Sigma \to T_{f(z)}M\,, \quad z \in \Sigma,$$

between Euclidean vector spaces, and therefore a norm

$$\| Tf \|^2 = \operatorname{tr}(Tf)^* \circ Tf.$$

The metric μ on Σ fixes a volume form $d\tau_\mu$. The *energy of* f is defined by

$$E(f) := \int_\Sigma \| Tf \|^2 \, d\tau_\mu$$

and the *area of* f by

$$a(f) = \int_\Sigma f^*(\omega).$$

The quoted lemma then says that for compatible (ω, J, g) we have

$$E(f) = 2a(f)$$

and that this is a topological invariant. Using the notation $[\omega]$ for the de Rham class of ω, this can also be written as $\langle[\omega], f_*(\Sigma)\rangle$. As a consequence, for example, one may derive that a pseudoholomorphic curve is a minimal surface relative to the metric.

Symplectic Capacities. In GROMOV [**Gr**] the following concept is also introduced and studied.

DEFINITION 2.33. The *symplectic radius*, $\operatorname{rad} M$, of a symplectic manifold M is the supremum of all r for which there is a symplectic embedding of the ball $\mathbf{B}(r)$ in M.

This motif was adopted and further expanded to the notion of the symplectic capacity by EKELAND, HOFER, VITERBO, ZEHNDER and others. This was a part of their study of the question of the existence of given closed curves, called *Hamiltonian trajectories* (to be defined in the next chapter), on the hypersurfaces of *constant energy* on symplectic manifolds. There is for this situation the following axiomatization (see AEBISCHER *et al.* ([**Ae**], pp. 73-74), or HOFER–ZEHNDER ([**HZ**], p. 51)).

DEFINITION 2.34. c is called a *symplectic capacity* of dimension $2n$ when c is a map which assigns to every symplectic $2n$–manifold (M, ω) (*eventually with boundary*) an element of $[0, \infty]$ satisfying the following properties:

C1 (*Monotonicity*): Whenever there exists a symplectic embedding $\phi : (M, \omega) \hookrightarrow (M', \omega')$, we have

$$c(M, \omega) \leq c(M', \omega').$$

C2 (*Conformality*): We have

$$c(M, \alpha\omega) = |\alpha| c(M, \omega) \quad \text{for all } \alpha \in \mathbb{R}^*.$$

C3 (*Non–triviality*): We have, for the standard form ω_0 on $\mathbb{R}^{2n}$,

$$c(\mathbf{B}(1), \omega_0) = c(\mathbf{Z}(1), \omega_0) = \pi.$$

The condition C3 can sometimes be weakened to

C3′ (*Weak non–triviality*): We have

$$0 < c(\mathbf{B}(1), \omega_0) \quad \text{and} \quad c(\mathbf{Z}(1), \omega_0) < \infty.$$

That a symplectic capacity actually is a symplectic invariant follows immediately from condition C1. These axioms, however, do not fix a uniquely defined capacity function.

Remark 2.35. 1) It can be shown that the square of the symplectic radius is a capacity.

2) For $n = 1$, thus in the case of a two–dimensional symplectic manifold, the total volume,

$$c(M, \omega) = \left| \int_M \omega \right|,$$

is an example of a capacity function, which for $(M, \omega) \subset (\mathbb{R}^2, \omega_0)$ agrees with the Lebesgue measure. For $n > 1$, $(\operatorname{vol} M)^{1/n}$ is not a capacity function, because of C3 and the fact that the cylinder has infinite volume.

EXERCISE 2.36. Show that, for an open subset $U \subset \mathbb{R}^{2n}$ and $\lambda \neq 0$ with the standard form ω_0, we have

$$c(\lambda U, \omega_0) = \lambda^2 c(U, \omega_0),$$

and calculate $c(\mathbf{B}(r), \omega_0)$.

With the help of the symplectic capacity, Gromov's squeezing theorem can be proven. For this and other applications, as well as for a proof of the existence of capacities, the reader is referred to HOFER–ZEHNDER ([**HZ**], in particular, Chapters 2–4), and to AEBISCHER *et al.* ([**Ae**], Chapter 4).

Chapter 3

Hamiltonian Vector Fields and the Poisson Bracket

Already in Sections 0.4 and 0.5 it was shown how the fundamental Hamilton's equations of theoretical mechanics can be elegantly formulated (and then also studied) with the help of the formalism of symplectic geometry. This formulation will be further detailed in this chapter. Sources for this material are, among others, KIRILLOV ([**Ki**], pp. 231–233), GUILLEMIN–STERNBERG ([**GS**], pp. 88 ff.) as well as ABRAHAM–MARSDEN ([**AM**], pp. 187–208).

3.1. Preliminaries

For a given vector field $X \in V(M)$, there are, as we've already partly seen in Chapter 2, several operations on the spaces of functions, vector fields and differential forms on M. You may find a collected, systematical overview on these topics in Sections A.4, A.5 and B.1. However, the information we now need is also collected here.

i) **Lie derivative of a function**. For $X \in V(M)$ and $f \in \mathcal{F}(M)$, we define $L_X f \in \mathcal{F}(M)$ by

$$L_X f\,(p) = df_p(X_p) \quad \text{for } p \in M;$$

thus

$$L_X f\,(p) = \sum_i a_i(x) \frac{\partial f}{\partial x_i}(x), \quad \text{when } X\big|_U = \sum a_i \frac{\partial}{\partial x_i}$$

in the chart (U, φ) with the coordinates x. Then L_X is an element of

$$\text{Der}\,(\mathcal{F}(M)) = \{D \in \text{End}\,\mathcal{F}(M); D(f \cdot g) = f\,Dg + g\,Df; f, g \in \mathcal{F}(M)\}.$$

ii) **Lie derivative of a differential form.** For $X \in V(M)$ and $\beta \in \Omega^k(M)$, we define $L_X\omega \in \Omega^k(M)$ by

$$L_X\beta = \frac{d}{dt}\,(F_t^*\beta)\,\big|_{t=0},$$

where F_t denotes the associated flow of X.

iii) **Lie derivative of a vector field.** For $X \in V(M)$ and $Y \in V(M)$, with F_t the flow associated to X, we define $L_XY \in V(M)$ by $L_XY = \frac{d}{dt}\,(F_{-t*}Y)|_{t=0}$, and it satisfies

$$L_XY = [X, Y].$$

Here we are using the fact that $V(M)$ is a Lie algebra with the Lie bracket $[\,,\,]$ as product.

iv) **Inner multiplication and the fundamental duality.** For $X \in V(M)$ and $\omega \in \Omega^2(M)$, the inner product of X and ω, $i(X)\omega$ (also written $\omega^b(X)) \in \Omega^1(M)$, is defined by

$$(i(X)\omega)(Y) = \omega(X, Y), \quad \text{for } Y \in V(M).$$

If ω is non–degenerate, this gives rise to an isomorphism

$$V(M) \underset{\omega^\#}{\overset{\omega^b}{\rightleftarrows}} \Omega^1(M).$$

In the standard symplectic coordinates $x = (q, p)$ of M we thus have $\omega = \sum_{j=1}^n dq_j \wedge dp_j$, and it follows that for

$$\vartheta = \sum_{j=1}^n (a'_j dq_j + a''_j dp_j) \in \Omega^1$$

the vector field $\omega^\#(\vartheta) = X \in V(M)$ corresponds to

$$X = \sum_{j=1}^n \left(a''_j \frac{\partial}{\partial q_j} - a'_j \frac{\partial}{\partial p_j}\right).$$

v) **A collection of formulas.** A few of the following statements have already been used previously. The reader may find proofs in Abraham–Marsden ([**AM**], pp. 109–120), but we also recommend that they be done as exercises. Let α be in Ω^k, $X, X_0, \ldots, X_k, Y$ elements of $V(M)$ and f an element of $\mathcal{F}(M)$. Then we have

1. $d\alpha\,(X_0,\dots,X_k) = \sum_{i=0}^{k}(-1)^i L_{X_i}(\alpha\,(X_0,\dots,\hat{X}_i,\dots,X_k))$
$$+\sum_{i<j}(-1)^{i+j}\alpha\big([X_i,X_j],X_0,\dots,\hat{X}_i,\dots,\hat{X}_j,\dots,X_k\big).$$

2. $i\,(X)\,\alpha$ is $\mathbb{R}$–bilinear in both X and α, and
$$i\,(fX)\,\alpha = fi(X)\,\alpha = i\,(X)f\alpha,\quad i\,(X)\,i\,(X)\,\alpha = 0,$$
$$i\,(X)(\alpha\wedge\beta) = (i\,(X)\,\alpha)\wedge\beta + (-1)^k\alpha\wedge(i\,(X)\beta)\qquad(\text{for } \beta\in\Omega^*).$$

3. $L_X\alpha = d(i\,(X)\,\alpha) + i\,(X)\,d\alpha.$

4. $L_X\alpha$ is $\mathbb{R}$–bilinear in both X and α, and
$$L_X(\alpha\wedge\beta) = L_X\alpha\wedge\beta + \alpha\wedge L_X\beta.$$

5. $(L_X\alpha)(X_1,\dots,X_k) = L_X\big(\alpha(X_1,\dots,X_k)\big)$
$$-\sum_{i=1}^{k}\alpha(X_1,\,\dots,\,[X,\,X_i],\dots,\,X_k).$$

6. $L_{fX}\alpha = fL_X\alpha + df\wedge(i\,(X)\,\alpha).$

7. $i\,([X,\,Y])\,\alpha = L_Xi\,(Y)\,\alpha - i\,(Y)L_X\alpha.$

8. $L_Xd\alpha = dL_X\alpha.$

9. $L_Xi\,(X)\,\alpha = i\,(X)L_X\alpha.$

10. $L_{[X,\,Y]}\alpha = L_XL_Y\alpha - L_YL_X\alpha.$

Remark 3.1. These formulas are dependent on the given formula for the wedge product of $\alpha_i\in\Omega^1(M)$ $(i=1,\dots,k)$ and $X_j\in V(M)$ $(j=1,\dots,k)$, which we have taken from ABRAHAM–MARSDEN. With this convention, we have the formula (see ABRAHAM–MARSDEN ([**AM**], p.104))
$$(\alpha_1\wedge\cdots\wedge\alpha_k)(X_1,\dots,X_k) = \sum_{\sigma\in\mathcal{S}_k}\operatorname{sgn}(\sigma)\alpha_1(X_{\sigma(1)})\cdot\cdots\cdot\alpha_k(X_{\sigma(k)}),$$
where the summation is taken over all the permutations σ of $1,\dots,k$. However, in the normalization that appears in, for example, KIRILLOV [**Ki**], one must take in the above formula (1) an additional factor of $k+1$ on the left side as well as applying an additional factor to the operator $i(X)$.

3.2. Hamiltonian systems

The following concept is central to the whole theory.

DEFINITION 3.2. Let $(M,\,\omega)$ be a symplectic manifold and $H \in \mathcal{F}(M)$. Then a vector field X_H on M is called a *Hamiltonian vector field* with the *energy function* H, if for X_H we have

$$i\,(X_H)\,\omega = dH.$$

$(M,\,\omega,\,X_H)$ is then called a *Hamiltonian system.*

The following important statements are, after the material of the previous section, but remarks.

Remark 3.3. If $(q,\,p)$ are the canonical coordinates of ω, then, by Section 3.1 iv), in these coordinates

$$X_H = \sum \left(\frac{\partial H}{\partial p_j}\frac{\partial}{\partial q_j} - \frac{\partial H}{\partial q_j}\frac{\partial}{\partial p_j} \right),$$

since $dH = \sum_j \left(\frac{\partial H}{\partial q_j} dq_j + \frac{\partial H}{\partial p_j} dp_j \right)$.

Remark 3.4. If $(M,\,\omega,\,X_H)$ is a Hamiltonian system, then there is an integral curve $\gamma = \gamma(t) = \big(q(t),\,p(t)\big)$, $t \in I$, for the vector field X_H, precisely when for this curve the *Hamiltonian equations* hold in their classical form,

$$\frac{\partial H}{\partial p_j} = \dot{q}_j, \quad \frac{\partial H}{\partial q_j} = -\dot{p}_j, \quad j = 1, \ldots, n.$$

And then the *theorem of the conservation of energy* has the form that $H(\gamma(t))$ is a constant for all $t \in I$.

Proof. $\gamma(t) = (q(t), p(t))$ is an integral curve for X_H precisely when

$$\dot{\gamma}(t) = (X_H)_{\gamma(t)} \quad \text{for all} \quad t.$$

With X_H as in the previous remark, this can be brought, with the use of Hamilton's equations, to the above form. And, when $\gamma = \gamma(t)$ is an integral curve to X_H, we have

$$\begin{aligned} \frac{d}{dt} H\left(\gamma\left(t\right)\right) &= dH_{\gamma(t)}(\dot{\gamma}(t)) = dH_{\gamma(t)}\big((X_H)_{\gamma(t)}\big) \\ &= \omega_{\gamma(t)}\Big((X_H)_{\gamma(t)}, (X_H)_{\gamma(t)}\Big) = 0. \end{aligned}$$

□

As an example of the usefulness of this Hamiltonian formalism we will show how to handle the example of the motion of a particle of mass m and charge e in an electromagnetic field of electric field strength $\mathbf{E} = (E_1, E_2, E_3)$ and magnetic field strength $\mathbf{B} = (B_1, B_2, B_3)$. Physics (for the appropriate physics we refer the reader to LANDAU–LIFSHITZ ([**LL**], p. 45)) then gives the equations of motion for these particles with velocity $\mathbf{v} = (v_1, v_2, v_3)$ and momentum $\mathbf{p} = (p_1, p_2, p_3)$ as

$$\frac{d\mathbf{p}}{dt} = e\mathbf{E} + \frac{e}{c}\,\mathbf{v} \times \mathbf{B},$$

and so

$$\frac{dp_i}{dt} = e\;E_i + \frac{e}{c}\,(v_j\,B_k - v_k\,B_j),$$

for each of the triplets $(i, j, k) \in \{(1, 2, 3),\ (2, 3, 1),\ (3, 1, 2)\}$. In this situation these quantities satisfy the relation

$$\mathbf{p} = \frac{m}{\sqrt{1 - v^2/c^2}}\,\mathbf{v}, \quad v^2 = \sum_{j=1}^{3} v_j^2\,,$$

which, in the classical case of $v^2 \ll c^2$, has the approximation $\mathbf{p} = m\mathbf{v}$. These equations allow one to find integral curves γ of a Hamiltonian system $(M,\,\omega,\,X_H)$.

EXERCISE 3.5. It is recommended that the reader determine the integral curves for the following systems (for the formulation of Maxwell's equations with differential forms see, for example, SCHOTTENLOHER ([**Sch**], p. 191).

a) In the classical case, for a time independent field, we have

$$M = T^*\mathbb{R}^3 \text{ with the coordinates } (q,\,p) = (q_1,\,q_2,\,q_3,\,p_1,\,p_2,\,p_3),$$

$$\omega = \omega_0 - (e/c)\omega_B \text{ with } \omega_0 = \sum_{j=1}^{3} dq_i \wedge dp_i,$$

$$\omega_B = \sum_{\{i,j,k\}} B_i dq_j \wedge dq_k,$$

$$H(q,\,p) = \frac{1}{2m}\sum_{i=1}^{3}{}' p_i^2 + e\phi(q) \text{ with } -\mathrm{grad}\phi = \mathbf{E}.$$

b) In the relativistic case, there is the *Minkowski space* $\mathbb{R}^{1,3}$ with the coordinates $q = (q_0,\,q_1,\,q_2,\,q_3)$ and the metric $ds^2 = dq_0^2 - dq_1^2 - dq_2^2 - dq_3^2$, so

$$M = T^*\mathbb{R}^{1,3} \text{ with the coordinates } (q,\, p),$$
$$\omega = c\widetilde{\omega}_0 + e\,\omega_F \text{ with}$$
$$\widetilde{\omega}_0 = dq_0 \wedge dp_0 - \sum_{i=0}^{3} dq_i \wedge dp_i,$$
$$\omega_F = \omega_B + \omega_E,$$
$$\omega_E = \sum_{i=1}^{3} E_i\, dq_i \wedge dq_0,$$

.

$$H(q,\, p) = \widetilde{H}(q,\, p) := (1/(2m))(p_0^2 - p_1^2 - p_2^2 - p_3^2).$$

Here one looks for curves $\gamma = \gamma(s) = (q(s),\, p(s))$ satisfying

$$\frac{d\gamma}{ds}(s) = (X_H)_{\gamma(s)}.$$

Hint. Use the following formula from relativity:

$$\frac{dt}{ds} = \frac{1}{c\sqrt{1 - v^2/c^2}}.$$

c) A variant of b) arises for the case that

$$M = T^*\mathbb{R}^{1,3},$$
$$\omega = c\widetilde{\omega}_0, \text{ and}$$
$$H(q,\, p) = \widetilde{H}(q,\, p - (e/c)A(q)) \text{ with } A = (A_0, A_1, A_2, A_3),$$
$$\text{so that } d\alpha = \omega_F \text{ for } \alpha = \sum_{i=0}^{3} A_i\, dq_i.$$

It is easy to see that b) and c) describe equivalent problems. In GUILLEMIN–STERNBERG ([**GS**], pp. 143-144) a recommendation is given of how a particle with *spin* would be described in this formalism.

Now we fill in more details of the general theory.

Remark 3.6. (Liouville's Theorem) Let F_t be the flow of X_H. Then F_t is symplectic; that is, we have

$$F_t^*\omega = \omega,$$

and consequently F_t preserves the volume form τ_ω.

Proof. From Lemma 2.3, we have

$$\frac{d}{dt}(F_t^*\omega) = F_t^*\Big(i\,(X_H)d\omega + d\big(i\,(X_H)\,\omega\big)\Big).$$

Thus, since $d\omega = 0$ and $i\,(X_H)\,\omega = dH$,

$$\frac{d}{dt}(F_t^*\omega) = F_t^*(ddH) = 0.$$

From this, $F_t^*\omega$ is seen to be independent of t, and since $F_0 = \mathrm{id}$, it must in fact be ω. From Section 1.1, the volume form of the given ω is fixed by

$$\tau_\omega = \frac{(-1)^{[n/2]}}{n!}\omega^n.$$

Since F_t preserves the 2–form ω, it must also preserve τ_ω. □

Since the Hamiltonian vector fields have such nice properties, we are compelled to give a special symbolism for them. Thus we let $\mathrm{Ham}(M)$ denote the *vector space of all Hamiltonian vector fields* and $\mathrm{Ham}^\circ(M)$ denote the *local Hamiltonian vector fields* $X \in V(M)$ that have the property that to every $m \in M$ there is a neighborhood U of m with $X\big|_U \in \mathrm{Ham}(U)$. The question of the characterization of Hamiltonian vector fields has the following answer.

THEOREM 3.7. *X is an element of* $\mathrm{Ham}^\circ(M)$ *exactly when one of the following equivalent conditions is satisfied:*

i) *$i\,(X)\,\omega$ is closed, or*

ii) *$L_X\omega = 0$ holds, or*

iii) *the flow F_t of X consists of symplectic maps.*

Proof. That a local Hamiltonian vector field satisfies the conditions i) – iii) is clear from the previous remark. Poincaré's lemma says that a closed form is locally exact, and so for $i\,(X)\,\omega$ there is a local function H with $dH\big|_U = i\,(X)\,\omega\big|_U$.

From the definition in 3.1 ii) of the Lie derivative of a differential form, we see that $L_X\omega = 0$ is equivalent to ω being fixed by F_t. This, along with the proof of the first equation in Remark 3.6, shows that $i\,(X)\,\omega$ is closed. □

Because of relation (10) in 3.1,

$$L_{[X,Y]}\omega = L_X L_Y\omega - L_Y L_X\omega,$$

local Hamiltonian vector fields give rise to a Lie algebra in $V(M)$. Hamiltonian vector fields are naturally locally Hamiltonian. For the reverse statement, a topological condition, which assures that closed 1–forms are also globally exact, is useful. This says, in effect (see section C.3), that the first homology group, $H^1(M, \mathbb{R})$, is $\{0\}$, which is exactly the fact needed. In particular,

Remark 3.8. For M with $H^1(M,\,\mathbb{R})=\{0\}$ we have that

$$\mathrm{Ham}(M)=\mathrm{Ham}^\circ(M)$$

is a Lie subalgebra of the Lie algebra $V(M)$.

In general the codimension of Ham in Ham° is exactly

$$b_1(M)=\dim H^1(M,\,\mathbb{R}),$$

the *first Betti number* of M.

An example of a locally Hamiltonian vector field which is not Hamiltonian will now be given. We begin with a vector field X, defined on the 2–dimensional torus $\mathbf{T}=\mathbb{R}^2/\mathbb{Z}^2$, with the coordinates $(x,\,y)$, by

$$X_{(x,y)}=(a,\,b),\quad a,\,b\in\mathbb{R}\text{ with }a^2+b^2\neq 0.$$

Naturally, here we have

$$\omega=dx\wedge dy.$$

Then

$$i\,(X)\,\omega=\big(i\,(X)dx\big)\wedge dy-dx\wedge\big(i\,(X)dy\big)=ady-bdx$$

is closed. In other words, this says that X is locally Hamiltonian. But a locally Hamiltonian vector field with no zero points on a *compact* symplectic manifold cannot be Hamiltonian. Thus if $X=X_H$, the function H must have a critical point (thus, a maximum or a minimum) on the compact manifold; but then X would have a zero point.

Remark 3.9. Let i denote the identification of each real number c with the constant function on M taking the value c at each point. Let j assign to each element $h\in\mathcal{F}(M)$ the associated Hamiltonian vector field X_h. Then there is a sequence of $\mathbb{R}$ vector spaces

$$0\longrightarrow\mathbb{R}\stackrel{i}{\longrightarrow}\mathcal{F}(M)\stackrel{j}{\longrightarrow}\mathrm{Ham}\,(M)\longrightarrow 0.$$

The fundamental duality $\omega^\#$ now says that this sequence is *exact*; this follows precisely because it is the constant functions which give rise to the trivial vector field. Thus $\mathrm{Im}\,i=\ker\,j$. We will call this the *fundamental exact sequence* .

It is of great significance that this sequence is not only an exact sequence of vector spaces, but is also short exact with regard to the Lie algebra structure, which is defined on $\mathcal{F}\,(M)$ by the Poisson brackets (see the following section).

3.3. Poisson brackets

There are many ways to introduce the Poisson brackets, which then naturally lead to the same result for the Poisson bracket of two functions in canonical coordinates (at least, up to sign, depending on whether the result is given in terms of $\omega_0 = dq \wedge dp$ or $\omega_0 = dp \wedge dq$). Here, we will continue to follow ABRAHAM–MARSDEN ([**AM**], p. 191), where Poisson brackets for 1–forms are first introduced. For this we will fix the symplectic manifold (M, ω). The isomorphism given in Section 3.1 iv),

$$V(M) \underset{\omega^\#}{\overset{\omega^b}{\rightleftarrows}} \Omega^1(M),$$

will be abbreviated to

$$\omega^b(X) =: X^b \quad \text{for} \quad X \in V(M),$$

respectively,

$$\omega^\#(\vartheta) =: \vartheta^\# \quad \text{for} \quad \vartheta \in \Omega^1(M).$$

DEFINITION 3.10. For α, $\beta \in \Omega^1(M)$, the *Poisson bracket* of α and β is the 1–form

$$\{\alpha, \beta\} := -[\alpha^\#, \beta^\#]^b.$$

From this we get the commutative diagram

$$\begin{array}{ccc} V(M) \times V(M) & \xrightarrow{-[\,,\,]} & V(M) \\ \Big\downarrow{\scriptstyle \omega^b \times \omega^b} & & \Big\downarrow{\scriptstyle \omega^b} \\ \Omega^1(M) \times \Omega^1(M) & \xrightarrow{\{\,,\,\}} & \Omega^1(M) \end{array}$$

Since $V(M)$ has a Lie algebra structure with respect to the Lie bracket $[\ ,\ ]$, so does $\Omega^1(M)$ with respect to the Poisson bracket $\{\ ,\ \}$.

THEOREM 3.11. *For* α, $\beta \in \Omega^1(M)$, *we have*

$$\{\alpha, \beta\} = -L_{\alpha^\#}\beta + L_{\beta^\#}\alpha + d\big(i(\alpha^\#)\, i(\beta^\#)\, \omega\big).$$

Proof. Here we will use the calculus of the Lie derivatives (see Section 3.1). The starting point is formula (1) of Section 3.1 (this is the same as material in Section C.2):

$$\begin{aligned}(d\omega)(X, Y, Z) &= L_X\big(\omega(Y, Z)\big) + L_Y\big(\omega(Z, X)\big) + L_Z\big(\omega(X, Y)\big) \\ &\quad -\omega\big([X, Y], Z\big) - \omega\big([Y, Z], X\big) - \omega\big([Z, X], Y\big).\end{aligned}$$

For $X = \alpha^{\#}, Y = \beta^{\#}$, it follows from the observation $\omega(\alpha^{\#}, Z) = \alpha(Z)$ that

$$\begin{aligned} 0 &= L_{\alpha^{\#}}(\beta(Z)) - L_{\beta^{\#}}(\alpha(Z)) - L_Z\big(i(\alpha^{\#})\, i(\beta^{\#})\, \omega\big) \\ &\quad + \{\alpha, \beta\}(Z) + \alpha(L_{\beta^{\#}} Z) - \beta(L_{\alpha^{\#}} Z), \end{aligned}$$

and from this, with an application of the standard formula $L_X Y = [X, Y]$, as well as (1) and (5) from Section 3.1, we derive that

$$0 = (L_{\alpha^{\#}}\beta)(Z) - (L_{\beta^{\#}}\alpha)(Z) - d\big(i(\alpha^{\#})\, i(\beta^{\#})\, \omega\big)(Z) + \{\alpha, \beta\}(Z).$$

□

COROLLARY 3.12. *If both* α *and* $\beta \in \Omega^1(M)$ *are closed, then* $\{\alpha, \beta\}$ *is exact.*

Proof. Since for a closed 1–form γ we have, from (3) in Section 3.1,

$$L_X \gamma = d(i(X)\gamma),$$

Theorem 3.11 immediately gives the result. □

The closed and the exact 1–forms generate Lie subalgebras of $\Omega^1(M)$.

Since every function $f \in \mathcal{F}(M)$ brings along a 1–form df and via $\omega^{\#}$ also a vector field X_f, we may define Poisson brackets on functions.

DEFINITION 3.13. For $f, g \in \mathcal{F}(M)$ we take as *Poisson bracket* the function

$$\{f, g\} := -i(X_f)\, i(X_g)\, \omega,$$

thus

$$= \omega(X_f, X_g).$$

In KIRILLOV ([**Ki**], p. 232), the statement of the following theorem is taken for the definition of the Poisson bracket.

THEOREM 3.14. *For* $f, g \in \mathcal{F}(M)$ *we have*

$$\{f, g\} = -L_{X_f} g = L_{X_g} f.$$

Proof. On the basis of the definition of X_g, we have

$$i(X_g)\, \omega = dg$$

and on the basis of the definition of the Lie derivative

$$\begin{aligned} L_{X_f} g &= dg(X_f) = \omega(X_g, X_f) = i(X_f)\, i(X_g)\, \omega \\ &= -\omega(X_f, X_g) = -L_{X_g} f. \end{aligned}$$

□

COROLLARY 3.15. *For* $f_0 \in \mathcal{F}(M)$, $g \mapsto \{f_0, g\}$ *is a derivation.*

Proof. For $g,\, h \in \mathcal{F}(M)$ we clearly have

$$\begin{aligned}\{f_0,\, gh\} &= -L_{X_{f_0}}(gh) = -d\,(gh)(X_{f_0}) = -\big((dg)h + gdh\big)(X_{f_0}) \\ &= -(L_{X_{f_0}}g)h - (L_{X_{f_0}}h)g = \{f_0,\, g\}h + \{f_0,\, h\}g.\end{aligned}$$

□

COROLLARY 3.16. *The following statements are equivalent:*

(1) *f is constant on the curves of X_g,*

(2) *g is constant on the curves of X_f, and*

(3) $\{f,\, g\} = 0$.

Proof. Let F_t be the flow of the vector field X_f. Then on the basis of the definition of the Lie derivative of a function (see Section 3.1 i), and Sections A.4 and A.5) we get that

$$\frac{d}{dt}(g \circ F_t) = F_t^* \, L_{X_f} g.$$

EXERCISE 3.17. Prove this last equality.

Now from Theorem 3.14 we also have

$$\frac{d}{dt}(g \circ F_t) = -\{f,\, g\} \circ F_t.$$

Thus g is constant along F_t exactly when $\{f,\, g\} = 0$. Because of the anti-symmetry of $\{f,\, g\}$, we also get the equivalence of ii). □

Remark 3.18. The relation $\{H,\, H\} = 0$ then says that H remains constant on the curves of the tangent vector field X_H, and therefore leads to the *theorem of the conservation of energy.*

Finally, we can bring the Poisson brackets from Section 0.5 into the picture:

COROLLARY 3.19. *In canonical coordinates, i.e., with a chart in whose coordinates $(q,\, p)$ the symplectic form ω has the standard form $\omega_0 = dq \wedge dp$, we have, for functions $f,\, g \in \mathcal{F}(M)$,*

$$\{f,\, g\} = \sum_{i=1}^{n} \left(\frac{\partial f}{\partial q_i}\frac{\partial g}{\partial p_i} - \frac{\partial f}{\partial p_i}\frac{\partial g}{\partial q_i} \right).$$

Proof. From Theorem 3.14

$$\{f, g\} = L_{X_g} f = df\,(X_g),$$

and from Section 3.1 iv)

$$X_g = \omega^{\#}(dg) = \frac{\partial g}{\partial p}\frac{\partial}{\partial q} - \frac{\partial g}{\partial q}\frac{\partial}{\partial p},$$

and so

$$\{f, g\} = \left(\frac{\partial f}{\partial q}\,dq + \frac{\partial f}{\partial p}\,dp\right)\left(\frac{\partial g}{\partial p}\frac{\partial}{\partial q} - \frac{\partial g}{\partial q}\frac{\partial}{\partial p}\right) = \frac{\partial f}{\partial q}\frac{\partial g}{\partial p} - \frac{\partial f}{\partial p}\frac{\partial g}{\partial q}.$$

□

COROLLARY 3.20. *For a Hamiltonian vector field $X_H \in \mathrm{Ham}(M)$ with flow F_t we have*

$$\frac{d}{dt}(f \circ F_t) = \{f \circ F_t,\, H\} \quad \textit{for all } f \in \mathcal{F}(M).$$

Proof. From Theorem 3.14 we have

$$\{f \circ F_t,\, H\} = L_{X_H}(f \circ F_t),$$

and so

$$\{f \circ F_t,\, H\} = d\,(f \circ F_t)(X_H) = \frac{d}{dt}(f \circ F_t).$$

□

The connection between the Poisson bracket of 1–forms and that of functions is now at hand.

THEOREM 3.21. *For $f, g \in \mathcal{F}(M)$, we have*

$$d\,\{f, g\} = \{df, dg\}.$$

Proof. From Theorem 3.11

$$\begin{aligned}\{df, dg\} &= -L_{X_f} dg + L_{X_g} df + d\left(i\,(X_f)\,i\,(X_q)\,\omega\right)\\ &= d(-L_{X_f} g + L_{X_g} f + i\,(X_f)\,i\,(X_q)\,\omega),\end{aligned}$$

and from Theorem 3.14 we have

$$\{df, dg\} = d\left(i\,(X_f)\,i\,(X_g)\,\omega\right) = d\{f, g\}.$$

□

Since $\Omega^1(M)$ with the Poisson bracket is a Lie algebra, we can easily arrive at the following important fact.

THEOREM 3.22. *$\mathcal{F}(M)$, as $\mathbb{R}$ vector space with the Poisson bracket, has the structure of a Lie algebra.*

Proof. Since d and $\omega^{\#}$ are $\mathbb{R}$–linear, the map $f \mapsto X_f$ is also $\mathbb{R}$–linear and it follows that $\{f, g\} = -i(X_f)\, i(X_g)\, \omega$ is $\mathbb{R}$–bilinear. It is then clear that $\{f, f\} = 0$. It remains to verify the Jacobi identity. We begin with

$$\begin{aligned} \{f, \{g, h\}\} &= -L_{X_f}\{g, h\} &&= L_{X_f}(L_{X_g} h), \\ \{g, \{h, f\}\} &= L_{X_g}(L_{X_h} f) &&= -L_{X_g}(L_{X_f} h), \\ \{h, \{f, g\}\} &= L_{X_{\{f,g\}}} h. \end{aligned}$$

But from Theorem 3.21, we have

$$X_{\{f,g\}} = \big(d\{f, g\}\big)^{\#} = \{df, dg\}^{\#} = -\big[(df)^{\#}, (dg)^{\#}\big],$$

and therefore

$$X_{\{f,g\}} = -[X_f, X_g],$$

from which the claim follows. □

The reader should pay particular attention to this last relationship.

COROLLARY 3.23. *For $f, g \in \mathcal{F}(M)$, we have*

$$X_{\{f,g\}} = -[X_f, X_g],$$

and so the Hamiltonian vector fields Ham(M) *generate a Lie algebra. The fundamental exact sequence in Remark 3.9 is, with $-j(f) = -X_f$, of the form*

$$0 \to \mathbb{R} \xrightarrow{i} \mathcal{F}(M) \xrightarrow{-j} \mathrm{Ham}(M) \to 0,$$

and, therefore, also an exact sequence of Lie algebras.

In the context of this formalism, we can give a useful criterion for when a diffeomorphism of symplectic manifolds is symplectic.

THEOREM 3.24. (Jacobi, 1837) *Let (M, ω) and (M', ω') be symplectic manifolds and $F : M \to M'$ a diffeomorphism. Then F is symplectic exactly when for all $h \in \mathcal{F}(M')$ we have*

$$F_* X_{h \circ F} = X'_h.$$

Proof. We begin with

$$X'_h = \omega'^{\#}(dh) \quad \text{and} \quad X_{h\circ F} = \omega^{\#}\big(d\,(h \circ F)\big).$$

Thus, for all $Y' \in V(M')$,

$$\omega'(X'_h, Y') = dh\,(Y').$$

i) When F is symplectic, we also have, for all $Y \in V(M)$,

$$\omega'(F_* X_{h\circ F},\, F_* Y) = \omega\,(X_{h\circ F},\, Y) = d\,(h \circ F)(Y).$$

Then, with the usual transformation formalism,

$$\omega'(F_* X_{h\circ F},\, F_* Y) = dh\,(F_* Y) = \omega'(X'_h,\, F_* Y).$$

Since ω' is nondegenerate, it follows that $F_* X_{h\circ F} = X'_h$ is closed.

ii) For $h \in \mathcal{F}(M')$ we have

$$i(X_{h\circ F})\omega = d(h \circ F) = F^* dh = F^*(i(X'_h)\,\omega').$$

The precondition $F_* X_{h\circ F} = X'_h$, along with the general formula

$$F^*(i(X')\alpha) = i((F^{-1})_* X')F^*\alpha,$$

gives

$$i(X_{h\circ F})\omega = i(X_{h\circ F})F^*\omega'.$$

Since every X_m can be taken locally to be of the form $(X_{h\circ F})_m$ for an $h \in \mathcal{F}\,(M)$, we arrive at the claim $F^*\omega' = \omega$. □

Symplectic maps can be defined by the property that they preserve the symplectic structure. They can also be characterized as preserving the Poisson bracket:

THEOREM 3.25. *The diffeomorphism F from M to M' is symplectic precisely when F preserves the Poisson bracket of functions on M or the Poisson bracket of 1–forms on M. In other words, exactly when*

$$\begin{aligned} F^* : \mathcal{F}\,(M') &\to \mathcal{F}\,(M), \\ f &\mapsto f \circ F, \end{aligned}$$

or, respectively,

$$\begin{aligned} F^* : \Omega^1(M') &\to \Omega^1(M), \\ \vartheta &\mapsto F^*\vartheta, \end{aligned}$$

are Lie algebra homomorphisms with relation to the Poisson bracket.

Proof. For $f,\, g \in \mathcal{F}\,(M')$, we have, from Theorem 3.14,

$$\{f,\, g\} \circ F = L_{(F^{-1})_* X'_g}(f \circ F),$$

and

$$\{f \circ F,\, g \circ F\} = L_{X_{g\circ F}}(f \circ F)$$

as well. Thus, from Theorem 3.24, both are the same exactly when F is symplectic. We leave the proof of the corresponding statements for 1–forms as an exercise. □

Now we are able to say whether the coordinates of a chart are canonical:

Remark 3.26. Let (U, φ) be a chart with coordinates (q, p). Then this is a *symplectic chart*, that is, with

$$\omega_0 = \sum dq_i \wedge dp_i,$$

we have $\omega = \varphi^* \omega_0$ exactly when

$$\{q_i, q_j\} = \{p_i, p_j\} = 0 \quad \text{and} \quad \{q_i, p_j\} = \delta_{ij},$$

for $i, j = 1, \dots, n$.

EXERCISE 3.27. Prove the remark.

3.4. Contact manifolds

We have now seen how specifying a 2–form with particular properties on a manifold will endow this manifold with the structure of a symplectic manifold, and that it will then necessarily have even dimension. In a precisely analogous manner, specifying a 1–form will define a contact structure on a manifold, which then forces the dimension to be odd. These contact manifolds can be discussed in a theory parallel to that of symplectic manifolds; however, both can also be developed as special cases of the general theory of *presymplectic manifolds.* In any case, it will soon be clear that the physically most important examples of contact manifolds arise as hypersurfaces of constant energy in Hamiltonian systems or from the treatment of time–dependent Hamiltonian functions. For this reason, we will give a few highlights from the theory of contact manifolds as a finish to this chapter on Hamiltonian systems. We begin by defining the major concept of this material.

DEFINITION 3.28. Let M be a differentiable manifold with dim $M = 2n + k$, $k \geq 0$, supplied with a 2–form ω which has rank $2n$ everywhere. Then ω is called a *presymplectic form* and (M, ω) a *presymplectic manifold.*

For $k = 0$ we get the definition of symplectic manifold, and for $k = 1$ we get that M is a so-called *weak contact manifold.* Darboux's Theorem 2.6 for symplectic manifolds transfers without difficulty to the following statement about the normal form of ω.

THEOREM 3.29. *Let M be a $(2n+k)$–dimensional differentiable manifold and ω a closed form of rank $2n$. Then for every point $m \in M$ there is a chart (U, φ) containing m with coordinates*

$$(q_1, \dots, q_n, p_1, \dots, p_n, w_1, \dots, w_k),$$

so that in these coordinates $\omega\big|_U$ can be written as

$$\omega\big|_U = \sum_{j=1}^{n} dq_j \wedge dp_j.$$

ABRAHAM–MARSDEN ([**AM**], p. 372) give a proof by falling back on the proof of Darboux's theorem that we gave in Section 2.2. On the other hand, STERNBERG ([**St**], pp. 137–140) gives a direct proof. The concept of a so-called *weak* contact manifold can be made sharper as in the following definition.

DEFINITION 3.30. A differentiable $(2n+1)$–dimensional manifold M is called a *contact manifold*, if there is on M a 1–form ϑ with

$$\vartheta \wedge (d\vartheta)^n \neq 0 \text{ everywhere on } M.$$

ϑ is then called a *contact form*.

This notation is in agreement with that found in AEBISCHER *et al.* [**Ae**], VAISMAN [**V**] and BLAIR [**Bl**]; this last is a good place to begin one's study of this theory. In BLAIR's text one may also find (pp. 11–12) some thoughts as to the choice of vocabulary. ABRAHAM–MARSDEN [**AM**] call the above defined object an *exact* contact manifold, and use the term *contact manifold* for what is here called a *weak* contact manifold.

It is clear that with $\omega = d\vartheta$ contact manifolds are weak contact manifolds. Corresponding to Darboux's Theorem 2.6, we have here a statement about the normal form of a contact manifold.

THEOREM 3.31. *Let M be a $(2n+1)$–dimensional differentiable manifold with a contact form ϑ. Then there is, for every point $m \in M$, a chart (U, φ) containing m with coordinates $(q_1, \ldots, q_n, p_1, \ldots, p_n, w)$ such that*

$$\vartheta\big|_U = dw + \sum_{j=1}^{n} p_j dq_j.$$

Proof. Let $\omega = -d\vartheta$. Then by Theorem 3.29 there are a chart (U, φ) and coordinates $(q_1, \ldots, q_n, p_1, \ldots, p_n, w_1)$ such that in U

$$d(\vartheta - \sum_{j=1}^{n} p_j dq_j) = 0,$$

and so locally for $w = w(q_1, \ldots, w_1)$

$$\vartheta - \sum_{j=1}^{n} p_j dq_j = dw.$$

Then, since $\vartheta \wedge (d\vartheta)^n \neq 0$, $(q_1, \ldots, q_n, p_1, \ldots, p_n, w)$ are also usable as coordinates. □

The reader may find a more direct proof in AEBISCHER *et al.* ([**Ae**], pp. 168–171).

The study of contact manifolds will be enlightened by the introduction of a particular differential system; that is, by the introduction of a system of smoothly varying subspaces of the tangent spaces of a fixed dimension, as we have already seen in the discussion of Frobenius' theorem in Section 2.5. Alternatively, this may be given as particular subbundles of the tangent bundle TM.

DEFINITION 3.32. Let ω be an element of $\Omega^2(M)$. Then we call

$$R_\omega := \{(m, X_m) \in TM \mid\ i(X_m)\omega_m = 0\}$$

a *characteristic bundle* of ω. $X \in V(M)$ is called a *characteristic vector field* of ω if

$$i(X)\omega = 0.$$

Remark 3.33. When $\omega \in \Omega^2(M)$ has constant rank, then R_ω is a subbundle of TM, and its sections generate a differential system (the *differential system of the characteristic vector fields*). When ω is closed, R_ω is involutive.

Proof. The proof is not hard (see ABRAHAM–MARSDEN ([**AM**], p. 371)). Since for two characteristic vector fields X, Y, $[X, Y]$ is also characteristic, we get the result from the formulas in Section 3.1. □

Remark 3.34. In particular, when ω defines a weak contact structure on M, then R_ω is a vector bundle of rank 1, thus the *characteristic line bundle.*

In analogy to the association of a characteristic bundle to a 2–form one may also associate a bundle to a 1–form.

DEFINITION 3.35. Let ϑ be an element of $\Omega^1(M)$ and $\vartheta_m \neq 0$ for all $m \in M$. Then we call

$$R_\vartheta := \{(m, X_m) \in TM;\ \vartheta_m(X_m) = 0\}$$

the *characteristic bundle of* ϑ.

Remark 3.36. (M, ϑ) is a contact manifold precisely when $d\vartheta$ is non-degenerate on all the fibers of R_ϑ.

Proof. Since R_ϑ is a vector bundle of rank $2n$, ϑ is thus non–degenerate on R_ϑ precisely when the n–th exterior power $(d\vartheta)^n \neq 0$. And this is equivalent to saying that $\vartheta \wedge (d\vartheta)^n \neq 0$ on all of M. □

One should be careful about the word usage, which we have here taken from ABRAHAM–MARSDEN [**AM**]. In BLAIR [**Bl**], the differential system D of a section of R_ϑ is called a *contact distribution* and an element X of a one–dimensional complement of D in $V(M)$ a *characteristic vector field* with the contact structure given by ϑ. Such an X is then fixed by $\vartheta(X) = 1$ and $d\vartheta(X, Y) = 0$ for all $Y \in V(M)$. The following examples should make clear what may be confusing at this first glance.

EXAMPLE 3.37. Let $M = \mathbb{R}^{2n+1}$ come with a contact structure via

$$\vartheta = dw - \sum_{i=1}^{n} p_i dq_i.$$

Then $X = \partial_w$ is, in the above sense, a characteristic vector field of ϑ, and, for $\omega = d\vartheta = \sum dq_i \wedge dp_i$, we have

$$\vartheta(X) = 1 \text{ and } i(X)\omega = 0;$$

that is, X also spans a one–dimensional space of characteristic vector fields to ω in the sense of the first definition. Further, the contact distribution D is here spanned by

$$X_i := \partial_{q_i} + p_i\partial_w \text{ and } X_{n+i} := \partial_{p_i},\ i = 1, \dots, n,$$

and then, clearly, for $i = 1, \dots, n$,

$$\vartheta(X_i) = \vartheta(X_{n+i}) = 0.$$

EXAMPLE 3.38. Let S_e be a *regular energy surface* for a Hamiltonian system (M, ω, H), that is, a connected component of $H^{-1}(e)$ for a regular value e of H; in other words, for those in which $dH_m \neq 0$ for all $m \in H^{-1}(e)$. S_e is then a submanifold of M of codimension 1. And it is $(S_e, \iota^*\omega)$ for $\iota : S_e \hookrightarrow M$, a weak contact manifold. $X_H\big|_{S_e}$ is a characteristic vector field of $\iota^*\omega$ and induces the characteristic line bundle $R_{\iota^*\omega}$ of $\iota^*\omega$. To see this is not difficult. In particular, we have $i(X_H\big|_{S_e})\iota^*\omega = 0$ because of the relation, defined on X_H,

$$\omega_m((X_H)_m, \xi) = (dH)_m(\xi) = 0$$

for all $m \in S_e$ and $\xi \in T_m S_e \subset T_m M$.

EXAMPLE 3.39. In order to produce a true contact manifold, we sharpen the last example, and arrive at the physically particularly interesting case of

$$M = T^*Q \xrightarrow{\pi_Q} Q$$

of the phase space of the configuration space Q with $\omega = -d\vartheta_0$ (see Section 2.3) and

$$H = K + V \circ \pi_Q,$$

where V is a real potential function on Q and K is the kinetic energy associated to the Riemannian metric. Then $(S_e, i^*\vartheta_0)$ is a contact manifold, and it can be easily shown that $\vartheta_0 \wedge (d\vartheta_0)^n$ on S_e has no zeroes (see ABRAHAM-MARSDEN ([**AM**], p. 373)).

This last example can be formulated in more abstract terms with the help of a theorem.

THEOREM 3.40. *Let $i : S \hookrightarrow \mathbb{R}^{2n+2}$ be the immersion of a smooth hypersurface, where no tangent space of S meets the origin of $\mathbb{R}^{2n+2}$. Then S has a contact structure.*

And it will be given by $i^*\alpha$ for

$$\alpha = x_1 dx_2 - x_2 dx_1 + \ldots + x_{2n+1} dx_{2n+2} - x_{2n+2} dx_{2n+1},$$

when $x_1, \ldots, x_{2n+2}$ are the coordinates in $\mathbb{R}^{2n+2}$ (for a proof, see BLAIR [**Bl**], pp. 9–10)).

The next example takes the form of a recommended exercise.

EXERCISE 3.41. Give the 3–dimensional torus $T = \mathbb{R}^3/\mathbb{Z}^3$ a contact structure as well as an associated characteristic vector field so that the contact distribution D can be made explicit.

EXAMPLE 3.42. Closely related to Example 3.39 is the example, particularly interesting for the physicist, of the construction of a contact manifold arising from a time–dependent Hamiltonian function. Here, one is given a symplectic manifold (M, ω) and a *time direction* $\mathbb{R}$ on $\mathbb{R} \times M$. Then we let

$$\begin{aligned} \pi_2 : \mathbb{R} \times M &\longrightarrow M, \\ (t, m) &\longmapsto m, \end{aligned}$$

$\tilde{\omega} := \pi_2^* \omega$, and $\underline{t}$ be the vector field on $\mathbb{R} \times M$ given by

$$\underline{t}_{(s,m)} = (1, 0) \in T_s\mathbb{R} \times T_m M \simeq T_{(s,m)}(\mathbb{R} \times M) \quad \text{for} \quad (s, m) \in \mathbb{R} \times M.$$

Then the following statements are easy to see (see ABRAHAM–MARSDEN ([**AM**], p. 374)).

Remark 3.43.

i) $(\mathbb{R} \times M, \tilde{\omega})$ is a weak contact manifold.

ii) $R_{\tilde{\omega}}$ is generated from the vector field $\underline{t} \in V(\mathbb{R} \times M)$.

iii) If $\omega = d\vartheta$, then for $\tilde{\vartheta} := dt + \pi_2^*\vartheta$ also $\tilde{\omega} = d\tilde{\vartheta}$, and $(\mathbb{R} \times M, \tilde{\vartheta})$ is a contact manifold.

EXAMPLE 3.44. This last example can be varied by the inclusion of a time–dependent Hamiltonian function H. For this the following formalism is useful. Let

$$X : \mathbb{R} \times M \rightarrow TM$$

be a time–dependent vector field; that is, for every fixed $t \in \mathbb{R}$, a fixed vector field on M is given. Then X represents, via

$$\begin{aligned} \tilde{X} : \mathbb{R} \times M &\longrightarrow T(\mathbb{R} \times M), \\ (t, m) &\longmapsto ((t, m), (1, X_{(t,m)}), \end{aligned}$$

a vector field $\tilde{X} \in V(\mathbb{R} \times M)$, called the *suspension of* X. To an integral curve γ of X through $m \in M$ (that is,

$$\begin{aligned} \gamma : I &\longrightarrow M, \\ t &\longmapsto \gamma(t), \end{aligned}$$

with $\dot{\gamma}(t) = X_{(t,\gamma(t))}$ for all $t \in I$ and $\dot{\gamma}(0) = m$) there is associated the integral curve $\tilde{\gamma}$ of $\tilde{X}$ through $(0, m) \in \mathbb{R} \times M$ given by

$$\begin{aligned} \tilde{\gamma} : I &\longrightarrow \mathbb{R} \times M, \\ t &\longmapsto (t, \gamma(t)), \end{aligned}$$

with $\dot{\tilde{\gamma}}(t) = (1, \dot{\gamma}(t)) = (1, X_{(t,\gamma(t))})$ and $\dot{\tilde{\gamma}}(0) = (0, m)$. If we now have a symplectic manifold (M, ω) along with a time–dependent Hamiltonian function $H \in \mathcal{F}(\mathbb{R} \times M)$, then with

$$\begin{aligned} H_t : M &\longrightarrow \mathbb{R}, \\ m &\longmapsto H_t(m) = H(t, m), \end{aligned}$$

X_{H_t} is, as in the above, a time–dependent Hamiltonian vector field. At each $m \in M$ the vector is represented by $(X_{H_t})_m$, and $\tilde{X}_H$ is the associated suspension. If we now put

$$\omega_H := \tilde{\omega} + dH \wedge dt,$$

we get the following statement as a variant of Remark 3.43.

Remark 3.45.

i) $(\mathbb{R} \times M, \omega_H)$ is a weak contact manifold.

ii) $\tilde{X}_H$ generates the characteristic bundle R_{ω_H} which satisfies

$$i(\tilde{X}_H)\omega_H = 0 \quad \text{and} \quad i(\tilde{X}_H)dt = 1.$$

iii) For $\omega = d\vartheta$, $\vartheta_H := \pi_2^*\vartheta + Hdt$ and we have $\omega_H = d\vartheta_H$. In the case that $H + (\vartheta \circ \pi_2)(X_H)$ vanishes nowhere, we have that $(\mathbb{R} \times M, \vartheta_H)$ is a contact manifold.

Thus here the contact manifold carries physical information, and the conservation of energy, which in the case of time–independent Hamiltonian functions H can be written as $L_{X_H} H = 0$, is here

$$L_{\tilde{X}_H} H = \frac{\partial H}{\partial t}.$$

The phenomena described in the examples can in large part be found in the following influential viewpoint of WEINSTEIN which forms the starting point for further study (see AEBISCHER *et al.* ([**Ae**], p. 174)). Let (M, ω) be a symplectic manifold and $i : S \hookrightarrow M$ a hypersurface given as the vanishing set of $f \in \mathcal{F}(M)$ with $df\big|_S \neq 0$. Then all the multiples of the Hamiltonian vector field X_f are called the *characteristic line fields* on S, and from this

$$\mathcal{L}_S := \{cX_f;\ c \in \mathbb{R}\}.$$

S is said to be of *contact type* if there is 1–form ϑ on S with

$$d\vartheta = i^*\omega \ \text{ and } \ \vartheta(X) \neq 0 \ \text{ for all } \ X \in \mathcal{L}_S \backslash \{0\}.$$

In the context of the concepts presented at the beginning of this section, every cX_f is then a characteristic vector field to ϑ in the sense of BLAIR as well as a characteristic vector field to $d\vartheta$ in the sense of Definition 3.35. The contact structure on S is naturally not uniquely defined; given β, a closed 1–form on S with $(\vartheta+\beta)(X) \neq 0$, then $\vartheta+\beta$ also defines a contact structure on S.

We close the chapter with a nice result that shows just how tightly symplectic and contact manifolds are interwoven.

THEOREM 3.46. *A manifold with an orientable contact structure can be realized as a hypersurface of contact type in a symplectic manifold.*

A proof of this statement, as well as an entrance to this modern development, can be found in AEBISCHER *et al.* ([**Ae**], pp. 167–218).

Chapter 4

The Moment Map

A very helpful technique used in classical mechanics for the solution of complex problems consists of deriving *integrals* (that is, expressions which remain constant with the motion of the system) by analyzing the symmetry of the given system. The conservation of momentum and angular momentum in systems with invariance under respectively translations, rotations, is the most frequent example. In this, the following beautiful and at first glance rather abstract formalism has been crystalized from the influential work of Souriau, Kostant, Smale and Marsden. This is linked to the discussion in Section 2.5.

4.1. Definitions

For what follows, we fix (M, ω) to be a symplectic manifold, on which the Lie group G *operates symplectically* via ϕ; that is, for

$$\begin{array}{rcl} G \times M & \to & M, \\ (g, m) & \mapsto & gm = \phi_g(m), \end{array}$$

where all the ϕ_g, $g \in G$, are symplectic diffeomorphisms with $\phi_e(m) = m$ and $\phi_{gg'}(m) = \phi_g(\phi_{g'}(m))$. We let $\mathfrak{g}$ denote the Lie algebra of G and $\mathfrak{g}^*$ its dual. In this situation, $\mathfrak{g}$ can be realized as the tangent space T_eG to G at the element $e \in G$, and this is in turn identified with the left–invariant vector fields $V_l(G)$ on G. In the same manner, $\mathfrak{g}^*$ can be realized as the cotangent space T_e^*G, which in turn is the left–invariant 1–forms on G.

For $X \in \mathfrak{g}$, we denote by X_M the vector field on M which for all $m \in M$ is given by the rule

$$(X_M f)(m) := \frac{d}{dt} f\left(\phi_{\exp tX}(m)\right)\Big|_{t=0} \quad \text{for } f \in \mathcal{F}(M),$$

or, equivalently, by

$$(X_M)_m := \frac{d}{dt}(\phi_{\exp tX}(m))\big|_{t=0},$$

or, yet again, by

$$(X_M)_m := (\psi_m)_{*e}X \qquad \text{for } \psi_m : G \longrightarrow M,$$
$$g \longmapsto \psi_m(g) = gm.$$

EXERCISE 4.1. It is recommended that the reader, using the introductory material of Appendix A, verify that these three definitions coincide with one another, and also verify the formula

$$L_{X_M}\omega = 0.$$

X_M will be called the *infinitesimal generator* of the operation on M associated to X.

DEFINITION 4.2. A map $\Phi : M \to \mathfrak{g}^*$ is called a *moment map* for the group operation ϕ if for all $Y \in \mathfrak{g}$ we have

$$(*) \qquad d\,\widehat{\Phi}\,(Y) = i\,(Y_M)\,\omega$$

with

$$\begin{aligned} \widehat{\Phi}\,(Y) : M &\to \mathbb{R}, \\ m &\mapsto \Phi\,(m)(Y). \end{aligned}$$

Remark 4.3. Here we have what might at first sight appear to be confusing notation:

$$\widehat{\Phi}\,(Y)(m) = \Phi\,(m)(Y) \quad \text{for } m \in M,\ \ Y \in \mathfrak{g},$$

which says, in any case, that Φ is fixed by $\widehat{\Phi}$. With the abbreviation of X_f for the (Hamiltonian) vector field $(*)$ associated to $f \in \mathcal{F}(M)$, this is equivalent to

$$X_{\widehat{\Phi}(Y)} = Y_M.$$

NOTATION 4.4. $(M,\, \omega,\, \phi,\, \Phi)$ is also called a *Hamiltonian G–space.*

Remark 4.5. Not every symplectic group operation has an associated moment map. The existence of a moment map Φ for a given operation ϕ is guaranteed only when the local Hamiltonian vector fields for M are also Hamiltonian. Then Theorem 3.7 says that the symplectic diffeomorphisms ϕ_g produce local Hamiltonian vector fields Y_M, which then, under the precondition $\operatorname{Ham}^\circ(M) = \operatorname{Ham}(M)$, are of the form $Y_M = X_{\widehat{\Phi}(Y)}$.

Remark 4.6. Should Φ and Φ' both be moment maps for the group operation ϕ, then there is a $\mu \in \mathfrak{g}^*$ with

$$\Phi\,(m) - \Phi'(m) = \mu \ \text{ for all } m \in M.$$

The meaning of the moment maps will be made a little clearer in the following statement about their behavior.

THEOREM 4.7. *Let Φ be a moment map for the operation ϕ of G on M, and let $H \in \mathcal{F}(M)$ be invariant under this operation, that is,*

$$H(m) = H\big(\phi_g(m)\big) \quad \text{for all } m \in M \quad \text{and } g \in G.$$

Then Φ is an integral for the vector field X_H associated to H; that is, for the flow F_t, $t \in I$ associated to X_H, we have

$$\Phi\big(F_t(m)\big) = \Phi(m) \quad \text{for all } m \in M,\ t \in I.$$

Proof. For all $Y \in \mathfrak{g}$, we have

$$H\big(\phi_{\exp tY}(m)\big) = H(m),$$

since H is invariant. Here, after differentiating with respect to t and evaluating at $t = 0$, we get

$$(dH)_m\big((Y_M)_m\big) = 0,$$

and so

$$L_{Y_M} H = 0 \ \text{ for } Y_M = X_{\widehat{\Phi}(Y)}.$$

From Theorem 3.14, we then have

$$\{H, \widehat{\Phi}(Y)\} = 0,$$

and from Corollary 3.16, we have

$$\widehat{\Phi}(Y)\big(F_t(m)\big) = \widehat{\Phi}(Y)(m)$$

for every $Y \in \mathfrak{g}$. Thus we arrive at the claim. □

In the most important examples, the moment map has yet another nice equivariance property. In order to describe this we recall the definition of the coadjoint representation from Section 2.5 or D.3. The *adjoint representation* Ad is given by

$$\begin{aligned} G \times T_eG &\to T_eG, \\ (g, Y) &\mapsto \mathrm{Ad}(g)Y = \mathrm{Ad}_g Y \quad \text{with } \mathrm{Ad}_g = (\varrho_g^{-1}\lambda_g)_{*e}, \end{aligned}$$

and the *coadjoint representation* Ad^* by

$$\begin{aligned} G \times (T_eG)^* &\to (T_eG)^* = T_e^*G, \\ (g, \alpha) &\mapsto \mathrm{Ad}^*(g)\alpha = (\mathrm{Ad}(g^{-1}))^*\alpha =: \mathrm{Ad}^*_{g^{-1}}\alpha, \end{aligned}$$ [1]

from which we define

$$\begin{aligned} (\mathrm{Ad}(g))^* : (T_eG)^* &\to (T_eG)^*, \\ \alpha &\mapsto \{Y \mapsto \alpha(\mathrm{Ad}(g)Y)\}, \end{aligned}$$

[1]This usage is common in the literature of the moment map, and we will use it here also.

and so

$$(\mathrm{Ad}(g)^*\alpha)(Y) = \alpha\,(\mathrm{Ad}(g)Y).$$

DEFINITION 4.8. A moment map Φ on the group operation ϕ is called Ad^*–*equivariant*, when

$$\Phi\left(\phi_g(m)\right) = \mathrm{Ad}^*_{g^{-1}}\Phi\,(m) \quad \text{for all } m \in M \quad \text{and } g \in G;$$

that is, when, for all $g \in G$, the following diagram commutes:

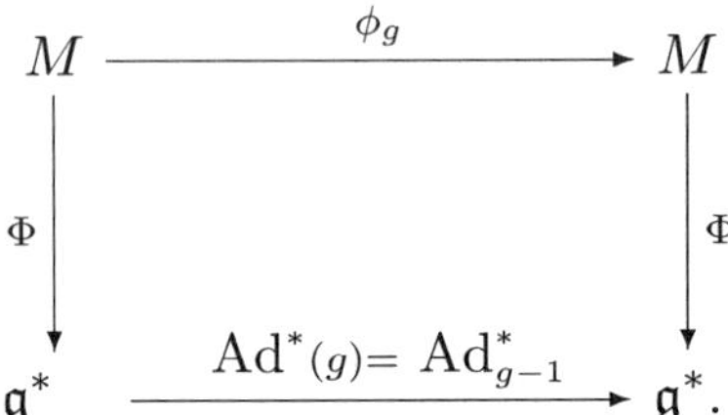

In ABRAHAM–MARSDEN [**AM**] the following is introduced as the means for weakening this equivariance.

DEFINITION 4.9. Let G be a Lie group and $\mathfrak{g}$ its Lie algebra. Then we call σ a *coadjoint cocycle* when σ is a map $\sigma : G \to \mathfrak{g}^*$ which satisfies the cocycle identity

$$\sigma\,(gh) = \sigma\,(g) + \mathrm{Ad}^*_{g^{-1}}\sigma\,(h) \quad \text{for all } g,\, h \in G.$$

Such a cocycle δ is called a *coboundary* if there is a $\mu \in \mathfrak{g}^*$ with

$$\delta\,(g) = \mu - \mathrm{Ad}^*_{g^{-1}}\mu \quad \text{for all } g \in G.$$

The cocycles modulo the coboundaries give a *cohomology group*, which is to be understood in connection with the general theory of Appendix C.

A Hamiltonian G–space $(M,\, \omega,\, \phi,\, \Phi)$ can be represented by a coadjoint cocycle σ (respectively, a cohomology class $[\sigma]$), in that for $g \in G$ and $\xi \in \mathfrak{g}$ we can take $\sigma(g)(\xi)$ as the value of the map $\psi_{g,\xi}$, which is defined by

$$\begin{aligned} \psi_{g,\xi} : M &\to \mathbb{R}, \\ m &\mapsto \widehat{\Phi}\,(\xi)(\phi_g(m)) - \widehat{\Phi}\,(\mathrm{Ad}_{g^{-1}}\xi)(m), \end{aligned}$$

and which can be shown to be constant (see ABRAHAM–MARSDEN ([**AM**], p. 277)).

4.2. Constructions and examples

We will now discuss only the Ad*–equivariant moment maps; we will try to explain the reason for the term *moment map*, in that the classical momentum and the angular momentum will both appear in the formalism. Before we can manage this, however, we must make a few general observations.

We proceed by taking as fixed a symplectic operation

$$\phi : G \times M \to M;$$

that is, for

$$G \times M \ni (g, m) \mapsto gm = \phi_g(m) \in M$$

all the ϕ_g, $g \in G$, are symplectomorphisms. For a $\xi \in \mathfrak{g} = \text{Lie}\, G$ we denote by ξ_M the associated *infinitesimal generator*, that is, the vector field on M which, as defined at the beginning of Section 4.1, is given by

$$(\xi_M)_m := \frac{d}{dt}\phi_{\exp t\xi}(m)\big|_{t=0}.$$

We will later need several properties of this generator:

Remark 4.10. For $\xi \in \mathfrak{g}$ and $g \in G$ we have

$$\begin{aligned}\big((\text{Ad}_g\xi)_M\big)_m &= (\phi_{g*})_{g^{-1}m}(\xi_M)_{g^{-1}m}\\ &= (\phi^*_{g^{-1}}\xi_M)(m)\end{aligned}$$

in the notation of ABRAHAM–MARSDEN ([**AM**], p. 269).

Proof. It follows from the definition of infinitesimal generators that

$$\big((\text{Ad}_g\xi)_M\big)_m = \frac{d}{dt}\phi_{\exp t\text{Ad}_g\xi}(m)\big|_{t=0},$$

and from the definition of Ad_g that

$$\big((\text{Ad}_g\xi)_M\big)_m = \frac{d}{dt}\phi_{g(\exp t\xi)g^{-1}}(m)\big|_{t=0}.$$

Then, since ϕ is a group operation,

$$\big((\text{Ad}_g\xi)_M\big)_m = \frac{d}{dt}\big(\phi_g \circ \phi_{\exp t\xi}(g^{-1}m)\big)\big|_{t=0},$$

which finally can be interpreted as the displacement of $(\xi_M)_{g^{-1}m}$ by ϕ_{g*} at the point $g\,(g^{-1}m) = m$. Thus

$$\big((\text{Ad}_g\xi)_M\big)_m = (\phi_{g*})_{g^{-1}m}(\xi_M)_{g^{-1}m}.$$

□

Remark 4.11. For $\xi, \eta \in \mathfrak{g}$ we have $[\xi_M, \eta_M] = -[\xi, \eta]_M$.

EXERCISE 4.12. Prove the remark. See ABRAHAM–MARSDEN ([**AM**], p. 269).

Remark 4.13. The notion of the infinitesimal generator is functorial. That is, when we are given two manifolds M and N with G–operations ϕ, respectively, ψ and an equivariant map $F : M \to N$ (thus with $F \circ \phi_g = \psi_g \circ F$ for all $g \in G$), then we have, for $\xi \in \mathfrak{g}$,

$$TF \circ \xi_M = \xi_N \circ F,$$

where ξ_M and ξ_N are the respective generators on M, respectively N. Thus we have the following commutative diagram:

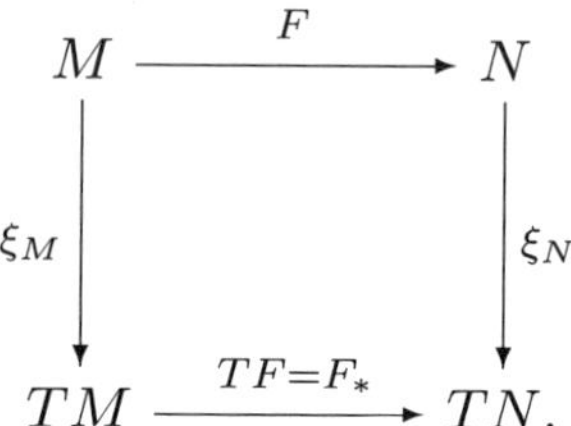

EXERCISE 4.14. Prove the remark. See ABRAHAM–MARSDEN ([**AM**], p. 270).

We recommend that readers take another opportunity to acquaint themselves with the notion of infinitesimal generators by proving the following.

EXERCISE 4.15. Let $\mathrm{Ad} : G \times T_eG \to T_eG$ be the adjoint representation. Then for $\xi \in T_eG \simeq \mathfrak{g}$ the associated infinitesimal generator is $\xi_{T_eG} =: \mathrm{ad}_\xi$ with

$$\begin{aligned} \mathrm{ad}_\xi : T_eG &\longrightarrow T_eG, \\ \eta &\longmapsto [\xi, \eta]. \end{aligned}$$

Many of the examples of moment maps stem from the following situation.

THEOREM 4.16. *Let ϕ be a symplectic operation of G on M. Let the symplectic form ω of M be exact (that is, $\omega = -d\vartheta$) and let the 1–form ϑ be G–invariant; thus $\phi_g^*\vartheta = \vartheta$ for all $g \in G$. Then*

$$\Phi : M \to \mathfrak{g}^*$$

with

$$\Phi\,(m)(\xi) = \big(i\,(\xi_M)\,\vartheta\big)(m) = \vartheta_m\big((\xi_M)_m\big)$$

defines an Ad^**–equivariant moment map for ϕ.*

Proof. i) Since ϑ is G–invariant, we have (compare with Section 3.1 ii))

$$L_{\xi_M}\vartheta = 0,$$

and because of (3) in Section 3.1 also

$$d(i(\xi_M)\,\vartheta) + i(\xi_M)d\vartheta = 0.$$

Thus

$$d\left(i(\xi_M)\vartheta\right) = i(\xi_M)\,\omega,$$

that is,

$$\widehat{\Phi}\,(\xi) := i(\xi_M)\vartheta \quad \text{for } \xi \in \mathfrak{g},$$

thus satisfying the characterizing relation of the moment map Φ.

ii) To demonstrate the Ad^*–equivariance, we must show that

$$\widehat{\Phi}\,(\xi)\big(\phi_g(m)\big) = \widehat{\Phi}\,(\mathrm{Ad}_{g^{-1}}\xi)(m) \quad \text{for all } g \in G,\ m \in M \text{ and } \xi \in \mathfrak{g}.$$

But from i) this is equivalent to

$$\left(i\,(\xi_M)\vartheta\right)\big(\phi_g(m)\big) = \Big(i\left((Ad_{g-1}\xi)_M\right)\vartheta\Big)(m),$$

and so to

$$\vartheta_{gm}(\xi_M)_{gm} = \vartheta_m\big((\mathrm{Ad}_{g^{-1}}\xi)_M\big)_m$$

and, because of Remark 4.10, even to

$$\vartheta_{gm}(\xi_M)_{gm} = \vartheta_m\big((\phi_{g^{-1}*})_{gm}(\xi_M)_{gm}\big).$$

Now the G–invariance of ϑ (that is, the property that $\phi_g^*\vartheta = \vartheta$) says that, for all $Y_m \in T_mG$, we have

$$\vartheta_{gm}\big((\phi_{g*})_m Y_m\big) = \vartheta_m(Y_m).$$

Here substituting $Y_m = \big((\mathrm{Ad}_{g^{-1}}\xi)_M\big)_m$ gives, on account of

$$(\phi_{g*})_m Y_m = (\phi_{g*})_m(\phi_{g^{-1}*})_{gm}(\xi_M)_{gm} = (\xi_M)_{gm},$$

the claimed equality

$$\vartheta_{gm}\big((\xi_M)_{gm}\big) = \vartheta_m\Big(\big((\mathrm{Ad}_{g^{-1}}\xi)_M\big)_m\Big).$$

□

This theorem will now be used on the *phase space*, i.e. the cotangent bundle $M = T^*Q$ of the *configuration space* Q. In this situation G operates via φ diffeomorphically on Q; that is, we have

$$\begin{array}{rcl} G \times Q & \to & Q, \\ (g,\,q) & \mapsto & gq = \varphi_g(q),\ \ \varphi_g \text{ diffeomorphic for all } g \in G. \end{array}$$

In light of Section 2.3 this G–operation on Q can be carried forward to a symplectic G–operation $\hat{\varphi}$ (also called a *canonical transformation*) on $M = T^*Q$ according to

$$\begin{array}{rcl} G \times T^*Q & \to & T^*Q, \\ (g,\,(q,\,\alpha_q)) & \mapsto & \big(gq = \varphi_g(q),\ \varphi^*_{g^{-1}}\alpha_q\big) =: \hat{\varphi}(q,\,\alpha_q), \end{array}$$

where the point $m \in M = T^*Q$ is written as the pair $m = (q, \alpha_q)$ with $q \in Q$ and $\alpha_q \in T^*_qQ$.

THEOREM 4.17. *$\hat{\varphi}$ has an* Ad**–equivariant moment map*

$$\Phi : M = T^*Q \to \mathfrak{g}^*.$$

For $m = (q, \alpha_q) \in M$ and $\xi \in \mathfrak{g}$ with infinitesimal generator ξ_Q on Q, thus,

$$(\xi_Q)_q = \frac{d}{dt}\varphi_{\exp t\xi}(q)\big|_{t=0},$$

Φ is given by

$$\widehat{\Phi}(\xi)(q, \alpha_q) := \alpha_q\big((\xi_Q)_q\big).$$

This can also be written as

$$\widehat{\Phi}(\xi) = P(\xi_Q),$$

where, for a vector field $X \in V(Q)$,

$$\begin{aligned} P(X) : \quad T^*Q &\to \mathbb{R}, \\ (q, \alpha_q) &\mapsto \alpha_q(X_q), \end{aligned}$$

is defined as the *momentum to* X.

Proof. The G–operation φ on Q is extended to a G–operation $\hat{\varphi}$ on $M = T^*Q$ so that the projection $\pi : T^*Q \to Q$ is G–equivariant; thus

$$\varphi_g \circ \pi = \pi \circ \hat{\varphi}_g.$$

The equivariance in the construction of the infinitesimal generators, as demonstrated in Remark 4.13, implies that, for $M = T^*Q$, $N = Q$ and $F = \pi$,

$$(*) \qquad \xi_Q \circ \pi = \pi_* \circ \xi_M.$$

The definition of the canonical 1–form ϑ on M (see Section 2.3), along with $d\pi_m = (\pi_*)_m : T_mM \to T_qQ$ for $m = (q, \alpha_q)$ and $X_m \in T_mM$, says that

$$\vartheta_{(q,\alpha_q)}(X_{(q,\alpha_q)}) = \alpha_q\left(\pi_{*(q,\alpha_q)}X_{(q,\alpha_q)}\right).$$

Employing this and $(*)$, we get

$$\begin{aligned} (i(\xi_M)\vartheta)(q, \alpha_q) &= \vartheta_{(q,\alpha_q)}\big((\xi_M)_{(q,\alpha_q)}\big) \\ &= \alpha_q\big((\pi_*)_{(q,\alpha_q)}(\xi_M)_{(q,\alpha_q)}\big) \\ &= \alpha_q\big((\xi_Q)_q\big) \\ &= P(\xi_Q)(q, \alpha_q). \end{aligned}$$

From this and Theorem 4.16, we can deduce that Φ is an Ad*–equivariant map. □

Were the usual coordinates $(q,\, p)$ for a point $m \in T^*Q$ used, then the momentum corresponding to $X_q = \sum \left(X_i(q)\, \partial_{q_i}\right) \in T_qQ$ would be

$$P\,(X)(q,\, p) = \sum_i p_i X_i(q).$$

In the treatment of quantization in the next section the following relations will receive meaning. These will be relations on liftings of functions $f \in \mathcal{F}\,(Q)$ as *position functions* to functions $\tilde{f} \in \mathcal{F}\,(M)$ such that $\tilde{f} = f \circ \pi$ for the projection

$$\begin{aligned} \pi : M = T^*Q \quad &\to \quad Q, \\ (q,\, \alpha_q) \quad &\mapsto \quad q. \end{aligned}$$

Remark 4.18. For $X,\, Y \in V\,(Q)$ and $f,\, g \in \mathcal{F}\,(Q)$ we have

i) $\{P\,(X),\, P\,(Y)\} = -P\left([X,\, Y]\right)$,

ii) $\{\tilde{f},\, \tilde{g}\} = 0$,

iii) $\{\tilde{f},\, P\,(X)\} = \widetilde{X\,(f)}$.

The proofs follow by routine computations (which appear in Abraham–Marsden ([**AM**], p. 284)).

From Theorem 4.17, a moment map can be established on the tangent bundle $M = TQ$ whenever Q is a Riemannian manifold with a scalar product $\langle\ ,\ \rangle$ on the tangent spaces T_qQ (it is actually sufficient that Q be pseudoriemannian) and a group G operates by isometries φ_g on Q such that this operation can be extended in a natural way to symplectomorphisms $\check{\varphi}_g = T\varphi_g$ on TQ. Then from Theorem 4.17, the following statement can be deduced (or it can be proven directly in exact analogy to the proof of that theorem).

Theorem 4.19. *The moment map Φ associated to $\check{\varphi}$ is given, for*

$$(q,\, v_q) \in TQ \quad \textit{with} \quad q \in Q,\; v_q \in T_qQ,$$

by

$$\Phi\,(q,\, v_q)(\xi) = \widehat{\Phi}\,(\xi)(q,\, v_q) := \langle v_q,\, (\xi_Q)_q\rangle \quad \textit{for} \quad \xi \in \mathfrak{g}.$$

Abraham–Marsden ([**AM**], p. 285) show that this is a special case of a general statement connected to Noether's theorem in the context of the Lagrangian formalism regarding TQ, which (pp. 208 ff.) is developed via the Hamiltonian formalism on T^*Q. On this topic we have only enough space to say a few words. A *Lagrange function* $L \in \mathcal{F}(TQ)$ can be assigned a *fiber derivative* $\mathbf{F}L$, that is, a map

$$\mathbf{F}L : TQ \to T^*Q,$$

which is defined by

$$(q,\, v_q) \mapsto (q,\, d''L_q),$$

where $d''L_q \in T_q^*Q$ at a $\tilde{v}_q \in T_qQ$ has the value $\big(d\,(L\big|_{\pi^{-1}(q)})\big)(\tilde{v}_q)$. (This corresponds to the relation $p = \frac{\partial L}{\partial \dot{q}}$ in Section 0.2 in the transition from the Lagrangian to the Hamiltonian formalism.) L is called *regular* when $\mathbf{F}L$ is a local diffeomorphism. Precisely in this case the symplectic standard form ω_0 on T^*Q is pulled back via $\mathbf{F}L$ to a symplectic form ω_L on TQ:

$$\omega_L = (\mathbf{F}L)^*\omega_0.$$

Correspondingly, the Liouville form ϑ carries over to

$$\vartheta_L := (\mathbf{F}L)^*\vartheta.$$

ABRAHAM–MARSDEN ([**AM**], pp. 285–286) now prove

THEOREM 4.20. *Let the regular Lagrangian function $L \in \mathcal{F}(Q)$ be G–invariant, that is,*

$$L \circ \check{\varphi}_g = L \quad \textit{for all } g \in G.$$

Then we have

i) *ϑ_L is also G–invariant; that is, $\check{\varphi}_g^*\vartheta_L = \vartheta_L$ for all $g \in G$.*

ii) *For this G–operation, an* Ad***–equivariant moment map Φ can be given by*

$$\widehat{\Phi}\,(\xi)(q,\, v_q) = d''L_q\big((\xi_Q)_q\big) \quad \textit{for } \xi \in \mathfrak{g}.$$

iii) *The moment map Φ is an integral for the L–associated Lagrange equation.*

Now we offer several examples.

EXAMPLE 4.21. Let $Q = \mathbb{R}^n$, and let $G = \mathbb{R}^n$ operate on $\mathbb{R}^n$ via translations

$$\begin{array}{rcl} G \times Q & \to & Q, \\ (s,\, q) & \mapsto & s + q = \varphi_s(q). \end{array}$$

Then the infinitesimal generator associated to $\xi \in \mathbb{R}^n = \mathfrak{g}$ is also $(\xi_Q)_q = \xi$ (and therefore independent of q). From Theorem 4.17 the associated moment map Φ on T^*Q is given in the standard coordinates $(q,\, p)$ of M by

$$\widehat{\Phi}(\xi)(q,\,p) = \sum_i p_i \xi_i,$$

which is also

$$\Phi(q,\,p) = p,$$

the *momentum*.

This is to be understood in connection with the conservation statement in Theorem 4.7. For every system with a Hamiltonian function invariant under the action of $G = \mathbb{R}^n$, the momentum is a conserved quantity (an observation which the physicist makes, however, without the need of the whole apparatus here constructed.)

EXAMPLE 4.22. Let $Q = \mathbb{R}^n$, and let G be a Lie subgroup of $GL_n(\mathbb{R})$. The elements $q \in Q$ will be thought of as columns, and G will operate as usual through multiplication; thus

$$\begin{aligned} G \times Q &\to Q, \\ (A,\,q) &\mapsto Aq = \varphi_A(q). \end{aligned}$$

The infinitesimal generator for $B \in \mathfrak{g} = \text{Lie } G \subset M_n(\mathbb{R})$ is B_Q with $(B_Q)_q = Bq$. We have further, from Theorem 4.17, that there is then an Ad^*–equivariant moment map given by

$$\widehat{\Phi}(B)(q,\,p) = p\,(Bq),$$

where p is meant to be a row vector.

In the special case of $n = 3$ and $G = SO(3)$, we have (see the end of Section B.2)

$$\mathfrak{g} = \mathfrak{so}\,(3) = \{B \in M_3(\mathbb{R}),\ B = -{}^tB\} \simeq \mathbb{R}^3$$

with

$$B = \begin{pmatrix} 0 & -b_3 & b_2 \\ b_3 & 0 & -b_1 \\ -b_2 & b_1 & 0 \end{pmatrix} \leftrightarrow b = \begin{pmatrix} b_1 \\ b_2 \\ b_3 \end{pmatrix} \in \mathbb{R}^3.$$

Were we now to realize the moment map Φ on TQ, as in Theorem 4.19, the result would be, with the standard scalar product $\langle\ ,\ \rangle$ on $\mathbb{R}^3$,

$$\begin{aligned} \widehat{\Phi}(B)(q,\,v) &= \langle v,\,Bq\rangle \\ &= \langle b \times q,\,v\rangle = \det(b,\,q,\,v) \\ &= \langle q \times v,\,b\rangle. \end{aligned}$$

Thus, with the identification of $\mathfrak{so}\,(3)$ with $\mathbb{R}^3$ and $\mathbb{R}^3$ with $(\mathbb{R}^3)^*$, we have

$$\Phi(q,\,v) = q \times v,$$

which is the usual *angular momentum*. For instance, for the harmonic oscillator with the Hamiltonian function

$$H\,(q,\,\dot{q}) = (1/2)(\|\,q\,\|^2 + \|\,\dot{q}\,\|^2),$$

this moment map is an integral.

EXAMPLE 4.23. The Lie group G operates on itself by left translation

$$\begin{array}{rcl} G \times G & \to & G, \\ (g,\,h) & \mapsto & gh = \lambda_g(h). \end{array}$$

Then the infinitesimal generator associated to this operation ξ_G for $\xi \in \mathfrak{g} =$ Lie G is given by the right–invariant vector field that takes on the value ξ at e; thus $(\xi_G)_g = (\varrho_{g*})_e\xi$, where ϱ_g is the right translation. From this we get for the moment map on T^*G

$$\widehat{\Phi}\,(\xi)(g, \alpha_g) = \alpha_g\big((\varrho_{g*})_e\xi\big) = (\varrho_g^*\alpha_g);$$

that is,

$$\Phi\,(g,\,\alpha_g) = \varrho_g^*\alpha_g(\xi).$$

ABRAHAM–MARSDEN [**AM**] discuss further examples in their exercises. Also related to the material of this section, GUILLEMIN–STERNBERG [**GS**] introduce the example of the operation of the Euclidean group $E\,(3) = SO(3) \ltimes \mathbb{R}^3$ on $\mathbb{R}^3$.

4.3. Reduction of phase spaces by the consideration of symmetry

A classical theorem, going all the way back to Jacobi and Liouville, says that by giving k *first* integrals whose Poisson brackets vanish, Hamilton's equations can be reduced to a system of equations in $2k$ fewer variables. In a similar way, rotational invariance in an n–body problem allows the elimination of four variables. These two processes indicate how, in a general way, with the help of *symplectic reduction*, one can reduce from higher to lower dimensional symplectic manifolds when a symmetry group operates on the given manifold. Going back to Élie Cartan, these procedures allow one to form *quotient spaces* and are a completely general central theme of later constructions; they have, in fact, already been seen at the end of Section 2.5 in the discussion of construction procedures for symplectic manifolds.

Here we will follow the treatment of ABRAHAM–MARSDEN ([**AM**], pp. 298 ff.). We assume that we are given

a symplectic manifold $(M,\,\omega)$,

a symplectic operation $\phi : G \times M \to M$ of a Lie group G on M, and for $\mathfrak{g} = \text{Lie } G$

an associated Ad^*–equivariant moment map $\Phi : M \to \mathfrak{g}^*$.

Then we denote by G_μ for a $\mu \in \mathfrak{g}^*$ the isotropy group

$$G_\mu := \{g \in G;\ \text{Ad}^*_{g^{-1}}\mu = \mu\}.$$

It is a general fact that this is a closed subgroup of G and therefore also a Lie group. Since Φ is Ad^*–equivariant, the space

$$M_\mu := \Phi^{-1}(\mu)/G_\mu$$

of G_μ–orbits on the fibers $\Phi^{-1}(\mu)$ makes sense. It is called the *reduced space* associated to the triple M, Φ and μ, and it is this *reduced phase space* which will find application in the important special case of $M = T^*Q$. We will begin by giving a few technical conditions that insure that M_μ is at least a smooth manifold.

i) Let $\mu \in \mathfrak{g}^*$ be a *regular* value for Φ; that is, for all $m \in \Phi^{-1}(\mu)$, the map $T\Phi_m$ (also written as $(\Phi_*)_m$) of the respective tangent spaces T_mM in $T_\mu\mathfrak{g}^* \simeq \mathfrak{g}$ is surjective (which, because of Sard's theorem, must be true for almost all μ). Then, with the methods described in Appendix A (see Abraham–Marsden [**AM**], p. 49), one may show that the fibers of $\Phi^{-1}(\mu)$ form a submanifold of M, and that this manifold has $\dim \Phi^{-1}(\mu) = \dim M - \dim G$.

ii) G_μ operates *without fixed points* and *properly* on $\Phi^{-1}(\mu)$. Here properly means the following: if (m_j) and $(\phi_{g_j} m_j)$ are convergent series in M, then $\hat{\phi}(g_j)$ has a convergent subsequence in G. This condition is, for example, automatically satisfied when G is compact. It is then a fundamental statement (see, for example, Abraham–Marsden ([**AM**], p. 266)) that the above introduced space $M_\mu = \Phi^{-1}(\mu)/G_\mu$ is a manifold and that the canonical projection

$$\pi_\mu : \Phi^{-1}(\mu) \to M_\mu = \Phi^{-1}(\mu)/G_\mu$$

is a submersion. That M_μ in this situation is symplectic is the content of the following theorem.

Theorem 4.24. *Let (M, ω) be symplectic with a symplectic G–operation and an* Ad^**–equivariant moment map satisfying the conditions given above. Then $M_\mu = \Phi^{-1}(\mu)/G_\mu$ has a uniquely defined symplectic form ω_μ with*

$$\pi_\mu^*\omega_\mu = i_\mu^*\omega,$$

where $\pi_\mu : \Phi^{-1}(\mu) \to M_\mu$ is the canonical projection and $i_\mu : \Phi^{-1}(\mu) \hookrightarrow M$ is the inclusion.

The proof requires the following statement.

LEMMA 4.25. *For $m \in \Phi^{-1}(\mu)$ and $Gm := \{\phi_g m;\ g \in G\}$ we have*

o) $T_m(\Phi^{-1}(\mu))/T_m(G_\mu m) \simeq T_{\pi_\mu(m)} M_\mu$,

i) $T_m(G_\mu m) = T_m(Gm) \cap T_m(\Phi^{-1}(\mu))$,

ii) $T_m(\Phi^{-1}(\mu))$ *and* $T_m(Gm)$ *are* ω*–orthogonal complements of one another.*

EXERCISE 4.26. Prove the lemma. In case of emergency, one may consult ABRAHAM–MARSDEN ([**AM**], p. 299).

Proof. Now the proof of the theorem follows in several steps.

a) For $v \in T_m\big(\Phi^{-1}(\mu)\big)$, let $[v] = (\pi_\mu)_{*m}(v)$ be the associated equivalence class in $T_m\big(\Phi^{-1}(\mu)\big)/T_m(G_\mu m)$. The equality $\pi_\mu^* \omega_\mu = i_\mu^* \omega$ says that

$$\omega_\mu\big([v],\, [w]\big) = \omega\,(v,\, w) \quad \text{for all } v,\, w \in T_m(\Phi^{-1}(\mu)).$$

Since π_μ and $(\pi_\mu)_*$ are surjective, ω_μ is clearly uniquely defined.

b) It follows immediately from part ii) of Lemma 4.25 that ω_μ is well-defined.

c) ω_μ is closed, and so

$$d\,(\pi_\mu^* \omega_\mu) = d\,(i_\mu^* \omega) = i_\mu^* d\omega = 0,$$

and so also $\pi_\mu^* d\omega_\mu = 0$. Then, because π_μ is surjective, we can conclude that $d\omega_\mu = 0$ is closed.

d) ω_μ is non–degenerate. Thus from

$$\omega_\mu\big([v],\, [w]\big) = 0 \quad \text{for all } w \in T_m \Phi^{-1}(\mu)$$

it follows that

$$\omega\,(v,\, w) = 0 \ \text{ for all } w \in T_m \Phi^{-1}(\mu),$$

thus $v \in T_m(Gm)$ from part ii) of the Lemma 4.25, and in this situation $v \in T_m(G_\mu m)$ from part i); that is, $[v] = 0$. □

It should be remarked that when $\omega = d\vartheta$ and ϑ is G–invariant, ω_μ need not be exact.

Remark 4.27. Since it is symplectic, the manifold M_μ has even dimension. Because of general principles that we cannot go into here, it turns out that

$$\dim M_\mu = \dim \Phi^{-1}(\mu) - \dim G_\mu = \dim M - \dim G - \dim G_\mu.$$

Remark 4.28. If μ is a regular value of Φ, the operation of G_μ is locally free. Then the reduction described in Theorem 4.24 can be taken, at least locally, in known interesting cases; this remains true even if the global conditions are not satisfied.

For the sake of completeness, we will now specialize the construction to the important case for physical applications of $M = T^*Q$. Here G operates on Q and then, as discussed in the previous section, also on M. The associated moment map is, as in Theorem 4.17, given by

$$\widehat{\Phi}\,(\xi)(q,\,\alpha_q) = \alpha_q\big((\xi_Q)_q\big) \quad \text{for } \xi \in \mathfrak{g},\ q \in Q,\ \alpha_q \in T_q^*Q.$$

Let the conditions for Theorem 4.24 be satisfied; then, moreover, G_μ operates without fixed points and properly on Q, so that

$$Q_\mu := Q/G_\mu$$

is again a manifold.

THEOREM 4.29. *Let α_μ be a G_μ–equivariant 1–form on Q with values in $\Phi^{-1}(\mu)$, that is, with $(\alpha_\mu)_q\big((\xi_Q)_q\big) = \mu\,(\xi)$ for all $\xi \in \mathfrak{g}$. Given the canonical symplectic 2–form ω_0 on T^*Q, the form*

$$\Omega_\mu := \omega_0 - \pi^* d\alpha_\mu$$

is then also a symplectic form on T^*Q. This then further induces a symplectic form on T^*Q_μ, and there is a symplectic embedding

$$\chi_\mu : M_\mu \to T^*Q_\mu$$

to a subbundle over Q_μ. χ_μ is then a diffeomorphism of T^*Q_μ precisely when $\mathfrak{g} = \mathfrak{g}_\mu = \operatorname{Lie} G_\mu$.

Sketch of proof. For

$$F_\mu := \{(q,\,\alpha_q) \in T^*Q;\ \alpha_q\big((\xi_Q)_q\big) = 0 \ \text{ for all } \xi \in \mathfrak{g}_\mu\}$$

we have, using, for example, results about quotients of bundles,

$$T^*Q_\mu = F_\mu/G_\mu.$$

From the definition of the fiber $\Phi^{-1}(\mu)$, we get (see Theorem 4.17)

$$\Phi^{-1}(\mu) = \{(q,\,\alpha_q) \in T^*Q; \alpha_q\big((\xi_Q)_q\big) = \mu\,(\xi) \text{ for all } \xi \in \mathfrak{g}\}.$$

Now define

$$\begin{aligned} \psi_\mu : \Phi^{-1}(\mu) &\to F_\mu, \\ (q,\,\alpha_q) &\mapsto \big(q,\,\alpha_q - (\alpha_\mu)_q\big), \end{aligned}$$

a symplectic map with $\psi_\mu^*\left(\Omega_\mu\big|_{F_\mu}\right) = \omega_0\big|_{\Phi^{-1}(\mu)}$:

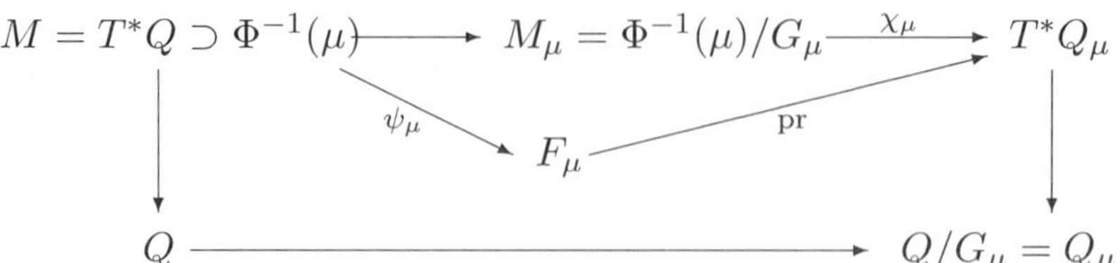

ψ_μ is clearly an embedding, and is surjective precisely when $\mathfrak{g} = \mathfrak{g}_\mu$. We also have that α_μ is G_μ–equivariant. With this $\mathrm{pr} \circ \psi_\mu$ factors through the quotients M_μ and so defines χ_μ. From the definition of symplectic structure on M_μ, χ_μ is symplectic. □

Remark 4.30. ABRAHAM–MARSDEN ([**AM**], Section 4.5) discuss more concrete systems with symmetry; these examples really bring Theorem 4.29 to life. They also construct differential forms α_μ of degree 1 for this theorem. It is worth remarking that for $\mu = 0$, one may assume that $\alpha_\mu = 0$. When G is abelian we have $\mathfrak{g} = \mathfrak{g}_\mu$, and so $M_\mu \simeq T^*Q_\mu$.

Examples.

EXAMPLE 4.31. The theorem of Jacobi and Liouville discussed at the beginning of this section now takes the following form. Let (M, ω) be a symplectic manifold, and let

$$f_1, \ldots, f_k \in \mathcal{F}(M) \text{ with } \{f_i, f_j\} = 0 \text{ for all } i, j.$$

Since the associated flows K_i, respectively K_j, to X_{f_i} and X_{f_j} commute (this is a consequence of Corollary 3.16), we get (locally) a symplectic operation of $G = \mathbb{R}^k$ defined on M. The associated moment map is then $\Phi = K_1 \times \ldots \times K_k$. It can then be taken that the df_i are independent at every point, so that every $\mu \in \operatorname{Im}\Phi$ is a regular value of Φ. Since G is abelian, we have $G_\mu = G$, and it gives rise to a symplectic manifold $M_\mu = \Phi^{-1}(\mu)/G$ of dimension $2n - 2k$.

It then remains to show (see ABRAHAM–MARSDEN ([**AM**], p. 304)) that every invariant Hamiltonian system on M canonically induces a Hamiltonian system on M_μ. In the important special case $n = k$, the system is called *completely integrable.*

In this sense, an example of a completely integrable system is the Hamiltonian system $(M = T^*\mathbb{R}^3,\ \omega_0,\ H_V)$ belonging to a *central force field*, thus with

$$H_V(q, p) = (1/2)p^2 + V(q),$$

where $V(q)$ is an $SO(3)$ invariant potential. Then

$$f_1 = H,\ f_2 = \| I \|^2,\ f_3 = I_3 \quad \text{with} \quad I = q \times p$$

are integrals with vanishing Poisson bracket.

EXERCISE 4.32. Verify this last statement.

EXAMPLE 4.33. Let X_H be a *Hamiltonian vector field* on M. Its flow gives a symplectic operation of $\mathbb{R}$ on M, whose moment map Φ is then the function H itself. Thus, for a regular value $E \in \mathbb{R}$ of H, we have a symplectic structure on $H^{-1}(E)/\mathbb{R} =: M_E$. In these quotients every orbit and every solution to the H–associated Hamilton's equations can be seen as a point; M_E is therefore called the *manifold for the solution of constant energy* E. We have $\dim M_E = \dim M - 2$.

EXAMPLE 4.34. For $G = SO(3)$, the adjoint operation of G on $\mathfrak{g} = \mathbb{R}^3$ can be written as in Example 4.22. For $\mu \in \mathbb{R}^3$, $\mu \neq 0$, $G_\mu = S^1$ then corresponds to the angular velocity of the μ–spanned straight line. The reduction of M to $\Phi^{-1}(\mu)/S^1 = M_\mu$ for the associated *angular momentum* Φ goes back to JACOBI. We have $\dim M_\mu = \dim M - \dim G - \dim G_\mu = \dim M - 4 = 2$.

EXAMPLE 4.35. Example 4.33 can be further extended to assign symplectic structures to specially given manifolds which can be considered as quotients in this scheme. This is the case for *complex projective space* $\mathbb{P}^n = \mathbb{P}^n(\mathbb{C})$ discussed in Section 2.6, and gives us a proof that this is a symplectic manifold. Let $M = \mathbb{R}^{2(n+1)} = T^*\mathbb{R}^{n+1}$ with the canonical symplectic form $\omega_0 = \sum_{i=1}^{n+1} dq_i \wedge dp_i$, and let

$$H(q, p) = \left(\frac{1}{2}\right) \sum_{i=1}^{n+1} (q_i^2 + p_i^2)$$

be the Hamiltonian function of the harmonic oscillator. Then we have (see Remark 3.3)

$$X_H(q, p) = \sum_{i=1}^{n+1} \left(p_i \frac{\partial}{\partial q_i} - q_i \frac{\partial}{\partial p_i} \right),$$

and the associated flow is

$$F_t : (q, p) \mapsto (q \cos t + p \sin t,\ p \cos t - q \sin t).$$

EXERCISE 4.36. Prove this last statement.

Since F_t is periodic with period 2π, F_t defines a symplectic operation of S^1 on M. Because of the compactness of S^1, the operation is proper and clearly fixed–point–free. The value $1/2$ is a regular value for H, and we have $H^{-1}(1/2) = S^{2n+1}$. Theorem 4.29, considered in the case of Example 4.33, then says that

$$H^{-1}(1/2)/\mathbb{R} = H^{-1}(1/2)/S^1 = S^{2n+1}/S^1 = \mathbb{P}^n(\mathbb{C})$$

is a symplectic manifold of real dimension $2n$.

EXAMPLE 4.37. We may also construct examples of symplectic manifolds with the help of the *coadjoint orbit* using the procedures discussed in Section 2.5, which can be, at least partly, adapted here. As in Example 4.23, let G be a Lie group, and $\lambda : G \times G \to G$ the action of G on itself by left translations; thus

$$(g, h) \mapsto \lambda_g(h).$$

$\widehat{\lambda}$ then denotes the usual extension of this operation on $M = T^*G$. Then, as shown in every example so far, the associated moment map Φ is given by

$$\Phi\,(g,\, \alpha_g) = \varrho_g^* \alpha_g.$$

Every $\mu \in \mathfrak{g}^*$ is a regular value for Φ, and we have

$$\Phi^{-1}(\mu) = \{(g,\, \alpha_g) \in T^*G, \;\; \alpha_g\big((\varrho_{g*})_g\, \xi\big) = \mu\,(\xi) \text{ for all } \xi \in \mathfrak{g}\};$$

that is, here α is the right invariant 1–form α_μ whose value at e is μ. Thus

$$\alpha = \alpha_\mu \text{ with } (\alpha_\mu)_e = \mu.$$

Moreover, it can be seen that

$$G_\mu = \{g \in G, \;\; \lambda_g^* \alpha_\mu = \alpha_\mu\}$$

and yet further that

$$\Phi^{-1}(\mu)/G_\mu \simeq G/G_\mu \simeq G^{\#}\mu \in \mathfrak{g}^*,$$

where the last isomorphism is given by the coadjoint representation. In this way the coadjoint orbit $G^{\#}\mu$ is seen to be a symplectic manifold. This is cited as the Kirillov–Kostant–Souriau theorem in ABRAHAM–MARSDEN ([**AM**], p. 302). They go on to explicitly construct the associated symplectic form ω_μ. The result which we gave as Theorem 2.27 then comes out.

EXERCISE 4.38. Let $M = \mathbb{C}^n$ be considered as a real symplectic manifold with $\omega(x, y) := \mathrm{Im}({}^t x \overline{y})$ for $x, y \in \mathbb{C}^n$. $G = S^1 = \{\zeta \in \mathbb{C}, \; |\zeta| = 1\}$ operates on M by multiplication:

$$(\zeta,\, x) \mapsto \Phi_\zeta(x) = \zeta x \;\text{ for } \zeta \in S^1, \;\; x \in M = \mathbb{C}^n.$$

Make explicit the above concepts in this simple situation and prove that ϕ_ζ is a symplectic operation, to which, for $x \in \mathbb{C}^n$ and $Y \in \mathfrak{g} = \mathrm{Lie}\; S^1 \simeq \mathbb{R}$,

$$\Phi(x)(Y) := (1/2) \parallel x \parallel^2 Y$$

gives a moment map. Are these Ad*–equivariant? What does the associated reduced symplectic space look like?

Here we stop so that we can save room for the last chapter, which will treat quantization. The reader interested in further concrete examples is unconditionally recommended to look over the remainder of Section 4.3, as well as Sections 4.4 and 4.5, in ABRAHAM–MARSDEN [**AM**]. This material should now be fairly easy to understand.

Chapter 5

Quantization

As we have already encountered in Section 0.6, the process of quantization converts the description of the time progression of physical systems on a symplectic manifold (in particular on that of the phase space) as described in classical mechanics to a similar progression on Hilbert space in quantum mechanics. The book by WALLACH [**Wa**] is dedicated to a critical account of these problems and includes many historical comments. Also well worth reading is the book by WOODHOUSE [**Wo**]. Yet another fully detailed account of these topics can be found in ABRAHAM-MARSDEN ([**AM**], Section 5.4, pp. 425 ff.). Shorter introductions occur in KIRILLOV ([**Ki**], Section 15.4) and in GUILLEMIN–STERNBERG ([**GS**], Section 34). However, these treatments require more supplementary materials from functional analysis and/or representation theory then we have covered in this treatise; a proper treatment of this preliminary material would carry us too far afield of the material of this text. The coverage of the general case in Section 5.5 is therefore, maybe, not based as soundly on the fundamentals as it could be. Fortunately the easiest conceivable situation, namely $M = \mathbb{R}^{2n} = T^*\mathbb{R}^n$, can be covered without a great deal of theoretical apparatus, and yet in it we can already see much of the problems and concepts coming into play. To make possible the introduction of the Heisenberg and Jacobi groups, which are equally important for physics and mathematics, we treat this simplest case in Sections 5.1 to 5.4 in quite some detail.

5.1. Homogeneous quadratic polynomials and $\mathfrak{sl}_2$

We begin by letting $M = \mathbb{R}^{2n}$ with the coordinates $(q_1, \ldots, q_n, p_1, \ldots, p_n)$. (However, when we really want to do some calculations, we will restrict to the case $n = 1$.) The starting point will be the fundamental exact sequence for Lie algebras (see Sections 3.2 and 3.3)

$$0 \to \mathbb{R} \xrightarrow{i} \mathcal{F}(M) \xrightarrow{-j} \operatorname{Ham}(M) \to 0.$$

Here i is the embedding which identifies $c \in \mathbb{R}$ with the constant function $f(m) = c$ for all $m \in M$, and j assigns to $H \in \mathcal{F}(M)$ the Hamiltonian vector field

$$X_H = \sum_{j=1}^{n} \left(\frac{\partial H}{\partial p_j} \frac{\partial}{\partial q_j} - \frac{\partial H}{\partial q_j} \frac{\partial}{\partial p_j} \right).$$

$\mathcal{F}(M)$ generates an (infinite–dimensional) Lie algebra with the Poisson bracket

$$\{f, g\} = \sum_{j=1}^{n} \left(\frac{\partial f}{\partial q_j} \frac{\partial g}{\partial p_j} - \frac{\partial f}{\partial p_j} \frac{\partial g}{\partial q_j} \right),$$

as does $\operatorname{Ham}(M)$ with the Lie bracket $[\ ,\ ]$, for vector fields. Thus we have

$$[X_f, X_g] = -X_{\{f,g\}} \quad \text{for } f,\ g \in \mathcal{F}(M).$$

Now in order to pass to the quantization, we search for an $\mathbb{R}$–linear map which assigns to elements of $\mathcal{F}(M)$ or, at a minimum, elements f from the largest possible subset of $\mathcal{F}(M)$, self-adjoint operators $\widehat{f}$ in a Hilbert space $\mathcal{H}$ such that the Lie structure is preserved in the sense that

$$(*) \qquad \widehat{\{f_1, f_2\}} = c\,[\widehat{f}_1, \widehat{f}_2] = c\,(\widehat{f}_1\widehat{f}_2 - \widehat{f}_2\widehat{f}_1),$$

where $c = -\frac{ih}{2\pi}$ (c is, up to the factor i, a factor ensuring the symmetry of the operators, a constant from physics, and h is called *Planck's constant*). Moreover, this map should extend

$$(**) \qquad \widehat{1} = 1 := \operatorname{id}_{\mathcal{H}}.$$

In view of the fact that M can be understood as $M = T^*\mathbb{R}^n$, we may take $\mathcal{H} = L^2(\mathbb{R}^n)$, which is here the *smallest non–trivial* Hilbert space. This search can now proceed with the following *general procedure.*

We look for a Poisson subalgebra $\mathcal{F}^0$ of $\mathcal{F}(M)$ which is isomorphic to the Lie algebra of a Lie group G,

$$\sigma : \mathcal{F}^0 \xrightarrow{\sim} \mathfrak{g} = \operatorname{Lie} G.$$

Then (see Appendix D) an *irreducible unitary representation* of G

$$\begin{aligned} \pi : G &\longrightarrow \operatorname{Aut}\mathcal{H}, \\ g &\longmapsto \pi(g), \end{aligned}$$

has an associated *infinitesimal representation* of $\mathfrak{g}$, which for $X \in \mathfrak{g}$ is given by

$$d\pi(X)v = \frac{d}{dt}\,\pi(\exp(tX)v)\big|_{t=0}.$$

This representation lives on the subspace $\mathcal{H}_\infty$ of *smooth* vectors v in $\mathcal{H}$, for which the differentiation process is practicable. There are many deep theorems from representation theory (in particular, from NELSON and HARISH–CHANDRA) which deal with this space and give answers to the question of when $\mathcal{H}_\infty$ is dense in $\mathcal{H}$. In the given case, $\mathcal{H}_\infty = \mathcal{S}(\mathbb{R}^n)$ is the Schwartz space, and all works well.

By differentiation of the unitarity condition

$$\langle \pi(\exp tX)v,\ \pi(\exp tX)w\rangle = \langle v,\ w\rangle$$

for $v, w \in \mathcal{H}_\infty$ and $\langle\ ,\ \rangle$ the scalar product in $\mathcal{H}$, we get

$$\langle d\pi(X)v,\ w\rangle + \langle v, d\pi(X)w\rangle = 0.$$

Thus the operators $d\pi(X)$ are *skew hermitian.* These become hermitian after multiplication by $\pm i$, and so, because

$$d\pi(\sigma(\{f,\ g\})) = [d\pi(\sigma(f)),\ d\pi(\sigma(g))],$$

the transformation

$$\mathcal{F}^0 \ni f \stackrel{\sigma}{\longmapsto} \sigma(f) = X \stackrel{d\pi}{\longmapsto} d\pi(X) \stackrel{\pm i}{\longmapsto} \pm i d\pi(X) =: \widehat{f}$$

satisfies the condition

$$\widehat{\{f,g\}} = \pm i d\pi(\sigma\{f,g\}) = \mp i[\widehat{f},\widehat{g}].$$

Choosing here the factor $-i$, we see that the condition $(*)$ is also satisfied; that is, once the constant $\frac{h}{2\pi}$ is normalized to 1 by an appropriate choice of units (as is often done in the case of physical problems). It is naturally easy here instead of multiplication by $-i$ to use multiplication by c^{-1}, so that the relation $(*)$ is realized with the constant c.

As the most obvious candidate for $\mathcal{F}^0$ to which the general procedure applies, we consider the Poisson algebra $\mathcal{F}_2$ of the quadratic polynomials in $\mathcal{F}$, thus (on account of the simplifying assumption $n = 1$)

$$\mathcal{F}^0 = \mathcal{F}_2 = \langle qp, p^2, q^2\rangle$$

with the relations

$$\begin{aligned} \{q^2, p^2\} &= 4qp, \\ \{qp, p^2\} &= 2p^2, \\ \{qp, q^2\} &= -2q^2. \end{aligned}$$

$\mathcal{F}_2$ is, via j, isomorphic to the subalgebra $\mathrm{Ham}_2(M)$ of $\mathrm{Ham}(M)$, which for

$$H_1 = qp, \quad H_2 = (1/2)p^2, \quad H_3 = (1/2)q^2$$

is generated by the vector fields

$$X_{H_1} = q\frac{\partial}{\partial q} - p\frac{\partial}{\partial p}, \quad X_{H_2} = p\frac{\partial}{\partial q}, \quad X_{H_3} = -q\frac{\partial}{\partial p}.$$

Thus $\mathcal{F}_2$ is isomorphic to the Lie algebra $\mathfrak{g} = \mathfrak{sl}_2 = \operatorname{Lie} SL_2(\mathbb{R})$, which can be written as

$$\mathfrak{sl}_2 = \langle H, F, G \rangle$$

with

$$H = \begin{pmatrix} 1 & 0 \\ 0 & -1 \end{pmatrix}, \quad F = \begin{pmatrix} 0 & 1 \\ 0 & 0 \end{pmatrix}, \quad G = \begin{pmatrix} 0 & 0 \\ 1 & 0 \end{pmatrix},$$

and so

$$[F, G] = H, \quad [H, F] = 2F, \quad [H, G] = -2G.$$

EXERCISE 5.1. It is recommended that the reader fill in the simple computational details for this example.

The reader may find discussions of the representations of $SL_2(\mathbb{R})$ and $\mathfrak{sl}_2$ in many sources (see in particular LANG: $SL_2(\mathbb{R})$ [**L3**], KNAPP [**Kn**], Chapter II, or BERNDT–SLODOWY [**BS**]). We will also give some details in Section 5.3.

EXERCISE 5.2. Let V be the vector space with basis $\{v_j | j \in \mathbb{N}_0\}$. For $\mu \in \mathbb{R}\backslash\{0\}$ let $\mathfrak{sl}_2$ act by

$$\begin{aligned} Hv_j &= (j + 1/2)v_j, \\ Fv_j &= -(1/(2\mu))v_{j+2}, \\ Gv_j &= (\mu/2)j(j-1)v_{j-2}. \end{aligned}$$

Verify that this describes a representation of $\mathfrak{sl}_2$.

It is worth bringing to the reader's attention that this representation is isomorphic to an infinitesimal representation of a unitary projective representation π_W of the group $SL_2(\mathbb{R})$, the so-called *Weil representation*, which after a little preparation will be covered in Section 5.3.

5.2. Polynomials of degree 1 and the Heisenberg group

The treatment above allows (at least, in principle) the quantization of systems consisting of homogeneous quadratic Hamiltonian functions in p and q. This should lead to, in particular, a description of the *harmonic oscillators*, but it is not quite enough. A little more can be accomplished by considering, instead, the case $\mathcal{F}^0 = \mathcal{F}_{\leq 2}(M)$ of polynomials of degree ≤ 2 in p, q. $\mathcal{F}_{\leq 2}(M)$ is also a Lie subalgebra relative to the Poisson bracket. As preparation for considering this case, we next study the case of $\mathcal{F}_1(M)$, which is the homogeneous linear polynomials in p, q. From the fact that

(1) $$\{q_i, p_j\} = \delta_{ij}, \ \{q_i, q_j\} = 0, \ \{p_i, p_j\} = 0, \quad i, j = 1, \ldots, n,$$

we see that $\mathcal{F}_1(M)$ is not a Lie subalgebra of $\mathcal{F}(M)$ (but $\mathcal{F}_{\leq 1}(M)$ is). The associated Hamiltonian vector fields are

$$X_p = \frac{\partial}{\partial q} \quad \text{and} \quad X_q = -\frac{\partial}{\partial p},$$

and so $\mathcal{F}_1(M)$ is bijectively carried by j to the *constant vector fields*

$$\mathrm{Ham}_c(M) = \mathbb{R}\frac{\partial}{\partial q} + \mathbb{R}\frac{\partial}{\partial p} \cong \mathbb{R}^{2n}.$$

$\mathrm{Ham}_c(M)$ is, in any case, a Lie subalgebra of $\mathrm{Ham}(M)$, which consists of the *infinitesimal translations* of $\mathbb{R}^{2n}$. We thus have before us the following situation: the Lie algebra $\mathfrak{g}_1 = \mathbb{R}^{2n}$ can be identified, via the map ι, with $\mathrm{Ham}_c(M)$. There is, however, a difference from the situation covered in Section 5.1, where $\mathfrak{sl}_2$ could be identified with $\mathcal{F}_2(M)$ and also with a subalgebra $\mathrm{Ham}_2(\mathcal{F})$ of $\mathrm{Ham}(\mathcal{F})$. Here there is no lifting of $\lambda : \mathfrak{g}_1 \to \mathrm{Ham}(M)$ to a Lie algebra homomorphism of $\mathfrak{g}_1$ to $\mathcal{F}(M)$ as in the following diagram:

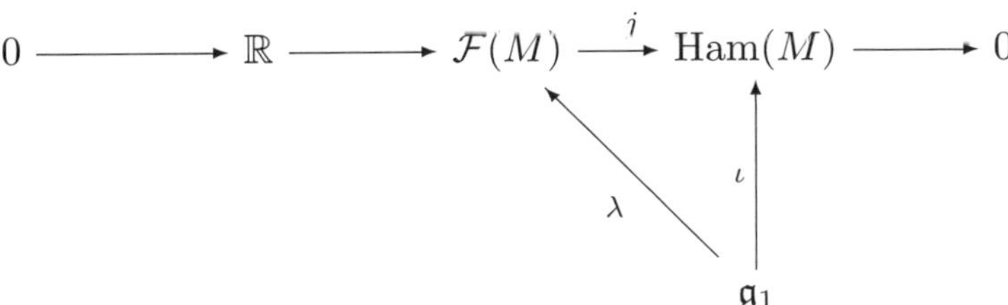

However, there is a Lie algebra, namely $\mathcal{F}_{\leq 1}(M)$, which is the image under j of $\mathrm{Ham}_c(M) \simeq \mathfrak{g}_1$; this is, however, not an isomorphism, but has a kernel $\mathrm{Ker}\, j \cong \mathbb{R}$. This situation gives us cause to look at the group $\mathbb{R}^{2n}$ of all translations of $\mathbb{R}^{2n}$, and to search for a group $H(\mathbb{R}^{2n})$ and an associated projection $\widehat{j}$ on $\mathbb{R}^{2n}$ so that for the associated Lie algebras $\widehat{j}$ is carried to j. Thus we are looking for $H(\mathbb{R}^{2n})$ and $\widehat{j}$ with

$$\begin{array}{ccc} H(\mathbb{R}^{2n}) & \xrightarrow{\ \widehat{j}\ } & \mathbb{R}^{2n} \\ \downarrow & & \downarrow \\ \mathfrak{h} \simeq \mathcal{F}_{\leq 1}(M) & \xrightarrow{\ j\ } & \mathfrak{g}_1 = \mathbb{R}^{2n}, \end{array}$$

that is, for a group satisfying the *Heisenberg commutation relations* (1). This task is solved by

The construction of the Heisenberg group.

Let (V, ω) be a symplectic $\mathbb{R}$ vector space of dimension $2n$. This gives rise to an associated Heisenberg group which is developed in various ways in the literature. We give a description of these.

a) The set

$$H(V) = V \times \mathbb{R} = \{h = (v, \kappa);\ \ v \in V, \kappa \in \mathbb{R}\}$$

is made into the Heisenberg group by the multiplication rule

$$h_1 h_2 = (v_1 + v_2, \kappa_1 + \kappa_2 + \omega(v_1, v_2)). \tag{3}$$

EXERCISE 5.3. Some simple calculations will show the following:

i) $H(V)$ is in fact a group. We have

$$h^{-1} = (-v, -\kappa)$$

and

$$h\, h_1 h^{-1} = (v_1,\ \kappa_1 + 2\omega(v,\ v_1)).$$

ii) $H(V)$ can be realized as the group of $(2n+2)$–rowed matrices. In particular, for $v = (\lambda, \mu)$ ($\lambda, \mu \in \mathbb{R}^n$ as rows), we have

$$h = \begin{pmatrix} 1_n & 0 & 0 & {}^t\mu \\ \lambda & 1 & \mu & \kappa \\ 0 & 0 & 1_n & -{}^t\lambda \\ 0 & 0 & 0 & 1 \end{pmatrix}.$$

iii) The one–parameter subgroups of $H(V)$ are of the form

$$G_{v,\kappa} := \{(tv, t\kappa), t \in \mathbb{R}\},\ \ v \in V,\ \kappa \in \mathbb{R}.$$

iv) For the associated Lie algebra $\mathfrak{h} = \mathfrak{h}(V)$ we have $\mathfrak{h} = V \times \mathbb{R}$ with

$$[(u_1,\ \vartheta_1), (u_2,\ \vartheta_2)] = (0,\ 2\omega(u_1, u_2))\ \text{ for }\ (u_1,\ \vartheta_1),\ (u_2,\ \vartheta_2) \in \mathfrak{h}. \tag{4}$$

v) For the adjoint representation of $H(V)$ on $\mathfrak{h}$ we have

$$\mathrm{Ad}_{(v,\kappa)}(u,\ \vartheta) = (u,\ \vartheta + 2\omega(v,\ u)).$$

vi) For $n = 1$, $\mathfrak{h} = \langle P, Q, R\rangle$ is with $[Q,\ R] = [P,\ R] = 0$ and $[P, Q] = 2R$,

$$P = \begin{pmatrix} 0 & 0 & 0 & 0 \\ 1 & 0 & 0 & 0 \\ 0 & 0 & 0 & -1 \\ 0 & 0 & 0 & 0 \end{pmatrix},\ Q = \begin{pmatrix} 0 & 0 & 0 & 1 \\ 0 & 0 & 1 & 0 \\ 0 & 0 & 0 & 0 \\ 0 & 0 & 0 & 0 \end{pmatrix},\ R = \begin{pmatrix} 0 & 0 & 0 & 0 \\ 0 & 0 & 0 & 1 \\ 0 & 0 & 0 & 0 \\ 0 & 0 & 0 & 0 \end{pmatrix}.$$

And now returning to the various constructions of the Heisenberg group:

b) For the second treatment, consider, for $S^1 = \{\zeta \in \mathbb{C},\ |\zeta| = 1\}$, the set

$$H'(V) = V \times S^1 = \{h' = (v, \zeta),\ v \in V,\ \zeta \in S^1\},$$

made into a group by

$$(3') \qquad h'_1 h'_2 = (v_1 + v_2,\ \zeta_1 \zeta_2 e(\omega(v_1, v_2))), \quad e(u) := \exp(2\pi i u).$$

Clearly we have an isomorphism of groups

$$H(V)/\mathbb{Z} \simeq H'(V)$$

with

$$h = (v,\ \kappa) \mapsto h' = (v,\ e(\kappa)).$$

c) GUILLEMIN–STERNBERG ([**GS**], p. 94) consider $V \times S^1$ with the product

$$(3^*) \qquad h_1 h_2 = (v_1 + v_2,\ \zeta_1 \zeta_2 \exp((i/2)\omega(v_1, v_2)))$$

This is written as $H^*(V)$.

d) Often (see for example the fundamental article of A. WEIL [**We1**] or the book of LION–VERGNE [**LV**]), given an arbitrary K vector space W with dual space W^* and associated canonical bilinear form $\langle\ ,\ \rangle : W \times W^* \to K$, one takes as the associated Heisenberg group

$$\text{Heis}\,(W) := W \times W^* \times K = \{h = (q,\ p,\ \kappa),\ q \in W,\ p \in W^*,\ \kappa \in K\}$$

with

$$h_1 h_2 = (q_1 + q_2,\ \ p_1 + p_2,\ \ \kappa_1 + \kappa_2 + \langle q_1, p_2 \rangle),$$

or

$$\text{Heis}'\,(W) := W \times W^* \times S^1 = \{h' = (q,\ p,\ \zeta);\ \ q \in W,\ \ p \in W^*,\ \zeta \in S^1\}$$

with

$$h'_1 h'_2 = (q_1 + q_2,\ \ p_1 + p_2,\ \ \zeta_1 \zeta_2 e(\langle q_1, p_2 \rangle)).$$

Here Heis (W) can be realized as a matrix group, specifically as a subgroup of $GL_{n+2}(K)$, via

$$h \longmapsto \begin{pmatrix} 1 & q & \kappa \\ 0 & 1_n & {}^t p \\ 0 & 0 & 1 \end{pmatrix}.$$

In what follows we will most often take as our starting point the description in a), which is also the one most often seen in the modern literature.

The Lie algebra $\mathfrak{h}$ of the Heisenberg group will naturally be called the *Heisenberg algebra*. For the natural basis elements $(n = 1)$

$$P = (1, 0, 0), \quad Q = (0, 1, 0), \quad R = (0, 0, 1)$$

of $\mathfrak{h}$ we have, on account of (4),

$$(5)\qquad [P,\, Q] = 2R, \quad [R,\, P] = [R,\, Q] = 0\ ,$$

and thus

$$\sigma(1) = 2R, \quad \sigma(q) = P, \quad \sigma(p) = Q$$

defines a Lie algebra isomorphism σ of $\mathcal{F}_{\leq 1}$ with $\mathfrak{h}$.

For the process of quantization, as formulated at the the beginning of Section 5.1, it is now interesting (although those who study representation theory do not need this additional motivation) to search for representations of $\mathfrak{h}$ and $H(V)$. This is also covered in the literature in a variety of ways. We begin by stating an analog to Exercise 5.1, that for $n = 1$ and $\mu \in \mathbb{R}\backslash\{0\}$, an irreducible representation of $\mathfrak{h} = \langle P,\, Q,\, R\rangle$ on the space W spanned by v_j, $j \in \mathbb{N}_0$, is given by

$$(6)\qquad Pv_j = v_{j+1}, \quad Qv_j = -\mu j v_{j-1} \quad , \; Rv_j = (\mu/2)v_j.$$

For quantization, this is not yet particularly helpful. It is important, however, that with just a little legwork a non–trivial unitary representation of the Heisenberg group can be found. And it can be fairly easily calculated that, for $h = (\lambda,\, \mu,\, \kappa) \in H(\mathbb{R}^{2n})$ and each $m \in \mathbb{R}$, a unitary representation of $H(\mathbb{R}^{2n})$ on $L^2(\mathbb{R}^n)$ can be given by

$$(7),\qquad (\pi_S^m(h)f)(x) = e^m(\kappa + (2x+\lambda)^t\mu)f(x+\lambda), \quad f \in L^2(\mathbb{R}^n),$$

which for $m \neq 0$ is called the *Schrödinger representation of index m.*

Exercise 5.4. Verify the necessary computations.

In the general theory this can be made to arise as a simple example of an *induced representation* in the sense of the theory of Mackey (see Appendix D; for more information, see, for example, Kirillov ([**Ki**], pp. 218-219), the already mentioned book by Lion–Vergne [**LV**] or the article of Cartier [**Ca**]). Guillemin–Sternberg ([**GS**], p. 97) give the following representation τ of $H^*(\mathbb{R}^{2n})$, which there arises as a result of a discussion on geometric optics(!). It is given with $h = (q,\, p,\, \zeta) \in H^*(\mathbb{R}^{2n})$ by the expression

$$(7')\qquad (\tau(h)f)(x) = \zeta^{-1}\, e^{-iq^tp}\, e^{ix^tp}\, f(x-q) \;\text{ for }\; f \in L^2(\mathbb{R}^n).$$

This corresponds to the representation π_S^m for $m = -1$ given above. The fundamental *theorem of Stone and von Neumann* says that such a representation π_S^m (respectively τ) of the Heisenberg group through its action on the center (that is, here through $\pi_S^m(0,0,\kappa) = e^m(\kappa)$) is, up to equivalence, uniquely defined. For a more precise formulation of this theorem and an accessible proof, the reader is referred to Lion–Vergne ([**LV**], pp. 19 ff.)

It is yet another easy exercise to calculate that from (7) it follows that (on account of simplicity we take $n = 1$)

$$\begin{aligned}\frac{d}{dt}\pi_S^m\left((t,0,0)\right)f\left(x\right)\Big|_{t=0} &= f'(x),\\ \frac{d}{dt}\pi_S^m\left((0,t,0)\right)f\left(x\right)\Big|_{t=0} &= 4\pi imxf(x),\\ \frac{d}{dt}\pi_S^m\left((0,0,t)\right)f\left(x\right)\Big|_{t=0} &= 2\pi imf(x).\end{aligned}$$

For the skew hermitian operators

$$d\pi_S^m(P) = \frac{d}{dx},\quad d\pi_S^m(Q) = 4\pi imx,\quad d\pi_S^m(R) = 2\pi im$$

the commutation rules (5) are satisfied. After multiplication by $-i$ (and normalization by $4\pi m = 1$) we get self-adjoint operators

$$(\hat{7})\qquad \hat{1} = 1,\quad \hat{q} = (1/i)\frac{d}{dx},\quad \hat{p} = x,$$

which is the first step in the process of quantization. Physicists may be uneasy that the position variable q represents a differential operator while p corresponds to a multiplication operator. But this follows from the particular form of the Schrödinger representation used here, which is the usual one from the standpoint of function theory. This can be easily converted to the form more familiar to physicists by considering instead a π_S^m–equivalent representation. The form we have chosen will very soon lead to the usual formula for the Weil representation, and so we will keep this possibly unfamiliar usage for now and make the appropriate transformation only in the larger context.

From the general theory (see Appendix D) it makes sense to carry this over to the complex numbers. In particular, let

$$Y_\pm := (1/2)(P \pm iQ),\quad Z_0 := -iR.$$

Then $\mathfrak{h}_c = \mathfrak{h} \otimes_{\mathbb{R}} \mathbb{C} = \langle Y_\pm, Z_0\rangle$ with

$$(5')\qquad [Z_0, Y_\pm] = 0 \quad\text{and}\quad [Y_+, Y_-] = Z_0,$$

as well as

$$(\hat{7}')\qquad \hat{\pi}_S^m(Y_\pm) = \frac{1}{2}\frac{d}{dx} \mp 2\pi mx \quad\text{and}\quad \hat{\pi}_S^m(Z_0) = 2\pi m.$$

Exercise 5.5. Verify that for $\mu > 0$ on $V = \langle v_j\rangle_{j\in\mathbb{N}_0}$ an irreducible representation of $\mathfrak{h}_c$ is given by

$$Y_+v_j = v_{j+1},\quad Y_-v_j = -j\mu\, v_{j-1},\quad Z_0v_j = \mu v_j,\quad j = 0,1,2,\ldots.$$

Show that this may be realized as functions on $L^2(\mathbb{R})$, beginning with

$$v_0 = f_0, \quad \text{with } f_0(x) = e^{-2\pi m x^2},$$

and then letting $Y_\pm$ and Z_0 act as differential operators, as given by $(\hat{7}')$, to specify v_j $(j \in \mathbb{N})$.

5.3. Polynomials of degree 2 and the Jacobi group

We will again focus, on account of its simplicity, on the case $n = 1$. For the most part, the following is not difficult to generalize to $n \geq 1$. Since $\mathcal{F}_{\leq 1}$ and $\mathcal{F}_2$ are Lie algebras, the subspace $\mathcal{F}_{\leq 2}$ of all polynomials of degree ≤ 2 is also a Lie algebra relative to the Poisson bracket. It is isomorphic to the Lie algebra

$$\mathfrak{g}^J = \mathfrak{sl}_2 + \mathfrak{h}$$

of the Jacobi group $G^J(\mathbb{R})$, which can be brought to light in the following way. The group $SL_2(\mathbb{R})$ operates on the right on $H(\mathbb{R})$ via

$$(M, h) \longmapsto h^M = (vM,\, \kappa) \text{ for } M \in SL_2(\mathbb{R}) \text{ and } h = (v,\, \kappa) \in H(\mathbb{R}).$$

This gives us justification to form the semidirect product

$$G^J(\mathbb{R}) := SL_2(\mathbb{R}) \ltimes H(\mathbb{R}) = \{g = (M,\, h) | M \in SL_2(\mathbb{R}),\, h \in H(\mathbb{R})\}$$

with the multiplication rule

$$gg' = (MM',\, vM' + v',\, \kappa + \kappa' + \omega(vM', v')).$$

EXERCISE 5.6. Verify the following:

a) We have $\mathfrak{g}^J = \langle H,\, F,\, G,\, P,\, Q,\, R \rangle$ with

$$\begin{aligned} &[F, G] = H, && [H,\, F] = 2F, && [H,\, G] = -2G, \\ &[P, Q] = 2R, && [H,\, Q] = Q, && [H,\, P] = -P, \\ & && [F,\, P] = -Q, && [G,\, Q] = -P \end{aligned}$$

and all the other Lie brackets are 0.

b) The map $\sigma : \mathcal{F}_{\leq 2} \longrightarrow \mathfrak{g}^J$ with

$$\begin{aligned} &\sigma(1) = 2R, && \sigma(q) = P, && \sigma(p) = Q, \\ &\sigma(qp) = H, && \sigma(q^2) = -2G, && \sigma(p^2) = 2F \end{aligned}$$

is a Lie algebra isomorphism.

Hint for a): $G^J(\mathbb{R})$ can be realized as the subgroup of $Sp_2(\mathbb{R})$ in which $H(\mathbb{R}^2)$ is embedded as described in Section 5.2 while introducing the Heisenberg group, and $SL_2(\mathbb{R})$ is embedded as follows:

$$M = \begin{pmatrix} a & b \\ c & d \end{pmatrix} \longmapsto \begin{pmatrix} a & 0 & b & 0 \\ 0 & 1 & 0 & 0 \\ c & 0 & d & 0 \\ 0 & 0 & 0 & 1 \end{pmatrix}.$$

Further details on this and for all that will follow in Section 5.3 can be found in BERNDT–SCHMIDT ([**BeS**], Chapters 1 and 2).

The operation of SL_2 on H fixes, in connection with the already mentioned theorem of Stone and von Neumann, a distinguished *projective* representation of SL_2, the so-called *Weil representation.* Using the Schrödinger representation, $\tau := \pi_S^m$ on $L^2(\mathbb{R})$, from Exercise 5.5, we can let every $M \in SL_2$ give a τ^M, which acts by

$$\tau^M(h) := \tau\left(h^M\right) \quad \text{for } h \in H,$$

giving a representation of H. Since this representation has the same effect as τ on the center of H, it follows from the theorem of Stone and von Neumann that they are equivalent; that is, for every $M \in SL_2$, there exists a unitary operator $U(M)$ such that

$$U(M)\,\tau(h)\,U(M^{-1}) = \tau(h^{M^{-1}}) \quad \text{for all } h \in H. \tag{8}$$

This relation fixes U up to a constant factor by a general statement from representation theory (the so-called *Schur's lemma*, see for example, KIRILLOV ([**Ki**], p. 119)). So we have

$$U(M_1)\,U(M_2) = c(M_1, M_2)\,U(M_1M_2). \tag{9}$$

Therefore $c(M_1, M_2) \in \mathbb{C}_1 = \{z \in \mathbb{C}, |z| = 1\}$, and for these c we have, on the basis of the associativity law in SL_2, the relation (see KIRILLOV ([**Ki**], p. 218))

$$c(M_1, M_2)\,c(M_1M_2, M_3) = c(M_1, M_2M_3)\,c(M_2, M_3).$$

This then says that

$$SL_2 \ni M \mapsto U(M) \in \operatorname{Aut} \mathcal{H} \tag{10}$$

defines a *projective representation* of SL_2 on $\mathcal{H}$.

Remark 5.7. A *projective representation* in the above sense $g \mapsto \pi(g)$ of a group G into the space $\mathcal{H}$ induces a *usual representation* $\widetilde{\pi}$ in the projective space $\mathbb{P}(\mathcal{H})$ associated to $\mathcal{H}$. This is defined by

$$v_\sim \mapsto \widetilde{\pi}(g)(v_\sim) = \big(\pi(g)\,v\big)_\sim \quad \text{for } v_\sim \in \mathbb{P}(\mathcal{H}).$$

This procedure accommodates well to quantum theory, since it is not the element v of the Hilbert space that has physical meaning, but the 1–dimensional subspace spanned by any such $v \neq 0$.

There is now a standard realization of (10), which is today most often called the *Weil representation*, although it also goes under the names of the Shale–Weil representation, the metaplectic representation or the oscillator representation. Also the name of Segal appears in this connection. This representation, here denoted by U instead of by π_W, is fixed by giving the actions of the generators of SL_2. This leads to the following formulas, which first appeared in WEIL's fundamental article [**We1**], which should be consulted since it introduces many other important concepts as well (also see, for example, LION–VERGNE ([**LV**], pp. 1–63), ([**Mu2**], pp. 133 f.), or GUILLEMIN–STERNBERG ([**GS**], pp. 47–65)). For $f \in L^2(\mathbb{R})$ we postulate

$$\begin{aligned}
\pi_W\big(d\,(a)\big)f\,(x) &= f(ax)|a|^{1/2} \\
&\quad \text{for } d\,(a) = \begin{pmatrix} a & 0 \\ 0 & a^{-1} \end{pmatrix}, \quad a \neq 0, \\
\pi_W\big(n\,(v)\big)f\,(x) &= f(x)e^{2\pi mivx^2} \\
&\quad \text{for } n\,(v) = \begin{pmatrix} 1 & v \\ 0 & 1 \end{pmatrix}, \qquad v \in \mathbb{R}, \\
\pi_W(w)f\,(x) &= \widehat{f}\,(x) = \int f(u)e^{2\pi iumx}du \\
&\quad \text{for } w = \begin{pmatrix} 0 & 1 \\ -1 & 0 \end{pmatrix}.
\end{aligned}$$

The derivation of these formulas needs more general considerations, but that they fix a representation can be verified by elementary means. Those who do not want to try this themselves may find the calculations in MUMFORD ([**Mu2**], pp. 134-135).

Together with the Schrödinger representation given in the previous section, $\pi_S^m = \tau$, we obtain via

$$\pi_{SW}(M,\, h) = \pi_S(h)\pi_W(M)$$

a (projective) representation of the Jacobi group $G^J(\mathbb{R})$, which we here call the *Schrödinger–Weil representation.*

EXERCISE 5.8. It is not such an easy calculation to determine the infinitesimal representation corresponding to π_{SW}. For the already given operators $(\hat{7})$ and $(\hat{7}')$ we have also

$$\text{(11)}\qquad \begin{aligned} \hat{\pi}_W(F) &= 2\pi i m x^2, \\ \hat{\pi}_W(G) &= \tfrac{i}{8\pi m}\left(\tfrac{d}{dx}\right)^2, \\ \hat{\pi}_W(H) &= \tfrac{1}{2} + x\,\tfrac{d}{dx} \end{aligned}$$

for

$$F = \begin{pmatrix} 0 & 1 \\ 0 & 0 \end{pmatrix}, \quad G = \begin{pmatrix} 0 & 0 \\ 1 & 0 \end{pmatrix}, \quad H = \begin{pmatrix} 1 & 0 \\ 0 & -1 \end{pmatrix},$$

respectively, for the complexification

$$X_\pm := (1/2)(H \pm i(F+G)), \ Z = -i(F-G),$$

$$\text{(11}'\text{)}\qquad \begin{aligned} \hat{\pi}_W(X_\pm) &= \tfrac{1}{4} + \tfrac{1}{2}\,x\tfrac{d}{dx} \mp \pi m x^2 \mp \tfrac{1}{16\pi m}\left(\tfrac{d}{dx}\right)^2, \\ \hat{\pi}_W(Z) &= 2\pi m x^2 - \tfrac{1}{8\pi m}\left(\tfrac{d}{dx}\right)^2. \end{aligned}$$

An extension of the last exercise allows us to obtain a realization of the infinitesimal representation $\hat{\pi}_{SW}$ on $V = \langle v_j \rangle_{j\in\mathbb{N}_0}$, where

$$\mathfrak{g}_c^J = \langle Z_0, Y_\pm, X_\pm, Z \rangle$$

for $\mu = 2\pi m > 0$ operates via

$$\begin{aligned} Z_0 v_j &= \mu v_j, & Z v_j &= (j+1/2)v_j, \\ Y_+ v_j &= v_{j+1}, & X_+ v_j &= -\tfrac{1}{2\mu}\,v_{j+2}, \\ Y_- v_j &= -\mu j v_{j-1}, & X_- v_j &= \tfrac{\mu}{2} j(j-1)v_{j-2}. \end{aligned}$$

The reader may find the necessary calculations in the notation used here in BERNDT–SCHMIDT ([**BeS**], Chapters 1 and 2) as well as, among others, LION–VERGNE ([**LV**], p. 198) and GUILLEMIN–STERNBERG ([**GS**], p. 101), where one must ever pay attention to the fact that the group law of the Heisenberg group has various formulations. In any case, we get here a representation of the Lie algebra relative to $\{\ ,\ \}$:

$$\mathcal{F}_{\le 2} = \langle 1, p, q, q^2, qp, p^2 \rangle$$

through the differential operators $(\hat{7})$ and (11), which operate on a dense subspace of $L^2(\mathbb{R})$ and after multiplication by $(-i)$ solve the quantization task from the beginning of Section 5.1.

5.4. The Groenewold–van Hove theorem

The investigations in Sections 5.1–5.3 dealt with the topic of quantization in the case of polynomials of degree 2, leading to the specification of the operators in (7) and (11). This then raises the question of whether this can be somewhat extended. The theorem of Groenewold and van Hove says, to a first approximation, that this is not possible. This will be made more precise in a moment; but first, we prove a statement of general interest for an arbitrary field K of characteristic 0.

THEOREM 5.9. *The algebra of the quadratic polynomials* $\mathfrak{p} = K[q, p]_{\leq 2}$ *is, relative to the Poisson bracket, a maximal subalgebra of the algebra* $K[q, p]$ *of all polynomials.*

Proof. i) The formulas

$$\{(1/2)q^2, p^j q^k\} = jp^{j-1}q^{k+1},$$

$$\{(1/2)p^2, p^j q^k\} = -kp^{j+1}q^{k-1},$$

$$\{pq, p^j q^k\} = (j-k)p^j q^k$$

show that by (possibly several–fold) left–Poisson bracket multiplication with q^2, respectively p^2, every monomial $p^j q^k$, $j + k = l$, in the set $\mathcal{F}_l$ of the homogeneous polynomials of degree l can be carried to any other; even more, that every homogeneous polynomial of degree l generates in this manner the whole of the monomials $p^j q^k$, $j + k = l$.

ii) Let $f \in K[q, p]$ be a polynomial of degree > 2. Because

$$\{q, f\} = \frac{\partial f}{\partial p} \quad \text{and} \quad \{p, f\} = -\frac{\partial f}{\partial q},$$

we can assume that, by taking left–bracket multiplication with p, respectively q, often enough, an algebra $\mathfrak{a}$ contained in the algebra $\mathfrak{p}$ has a polynomial with maximal homogeneous component of degree 3, and after subtraction of the quadratic part the polynomial is homogeneous of degree 3. From i), then, the whole of $\mathcal{F}_3$ lies in the algebra $\mathfrak{a}$.

iii) Because

$$\{q^3, p^3\} = 9q^2p^2$$

is a polynomial of degree 4, we have from i) that the whole of $\mathcal{F}_4$ is in the algebra. And now, since

$$\{q^3, p^n\} = (3n)q^2p^{n-1},$$

this can be extended inductively to show that all polynomials of degree higher than 2 similarly lie in $\mathfrak{a}$. $\square$

From $(\hat{7})$ in Section 5.2 as well as (11) in Section 5.3, after the normalization $4\pi m = 1$, the map σ :

$$\begin{aligned} 1 &\longmapsto i, \\ q &\longmapsto \frac{d}{dx}, \\ p &\longmapsto ix, \\ q^2 &\longmapsto -i\left(\frac{d}{dx}\right)^2, \\ p^2 &\longmapsto ix^2, \\ qp &\longmapsto \left(x\frac{d}{dx} + 1/2\right) \end{aligned}$$

is a representation of $\mathfrak{p} = \mathcal{F}_{\leq 2}$ which acts by skew hermitian operators on a dense subspace of $\mathcal{H} = L^2(\mathbb{R})$, and after multiplication by $(-i)$ can be made hermitian. This can be shown by direct calculation, without returning to the formulas for the Schrödinger–Weil representation.

EXERCISE 5.10. Verify the above construction, and verify the statements in the following construction.

To adapt our presentation to the one in GUILLEMIN–STERNBERG ([**GS**], pp. 101–104), which goes back to CHERNOFF and will be here varied a bit, and to simultaneously manufacture the properties most familiar to physicists, we introduce a Lie algebra isomorphism η of $\mathfrak{p}$ given by

$$\begin{array}{llllll} \eta(1) = -1, & \eta(q) = -p, & \eta(p) = -q, \\ \eta(q^2) = -p^2, & \eta(p^2) = -q^2, & \eta(qp) = -qp. \end{array}$$

Then $\sigma\eta$ is a Lie algebra isomorphism of $\mathfrak{p}$, and $A = i\sigma\eta$ is given by

$$\begin{aligned} A(1) &= 1, \\ A(q) &= x, \\ A(p) &= -i\,\tfrac{d}{dx}, \\ A(q^2) &= x^2, \\ A(p^2) &= -\left(\tfrac{d}{dx}\right)^2, \\ A(qp) &= -i(x\tfrac{d}{dx} + 1/2). \end{aligned} \tag{12}$$

Setting $\hat{f} = Af$, we see that $\sigma\eta$ satisfies the quantization conditions $(*)$ and $(**)$ given at the beginning, and so

$$\widehat{\{f,g\}} = -i[\widehat{f},\widehat{g}] \quad \text{and} \quad \widehat{1} = 1.$$

To simplify the notation, we now put

$$A(q) = x =: Q \quad \text{and} \quad A(p) = -i\,\frac{d}{dx} =: P$$

(the old meanings of the letters P and Q will no longer be needed). Then we clearly have

(13) $$A(q^2) = Q^2 \quad \text{and} \quad A(p^2) = P^2.$$

In this situation, Groenewold–van Hove's theorem is contained in the following:

THEOREM 5.11. *There are no linear maps A from $K[q, p]$ into an associative K algebra that satisfy $(*)$ and $(**)$ as well as (13) with $A(q) = Q$ and $A(p) = P$.*

This means in particular that the quantization map A cannot be extended to $\mathfrak{p}$, since every properly contained Poisson algebra of $\mathfrak{p}$ is by Theorem 5.9 equal to $K[q, p]$.

Proof. i) The basic idea of the proof is that, because of the plethora of relations satisfied by the Poisson brackets, the polynomials can be represented by a Poisson bracket in a variety of ways. In our case we particularly consider

(14) $$q^2p^2 = (1/3)\{q^2p, p^2q\}$$

and

(15) $$q^2p^2 = (1/9)\{q^3, p^3\}.$$

From these we will construct a contradiction which says that A assigns to the same Poisson bracket differing values. This proceeds by a few calculations and will be given in several steps, beginning with:

ii) We have

(16) $$A(qp) = (1/2)(QP + PQ).$$

In fact, from $\{q, p\} = 1$ it follows with $(*)$ that

(17) $$[Q, P] = QP - PQ = i$$

and from $\{q^2, p^2\} = 4qp$

$$4A(qp) = -i[Q^2, P^2] = -i(Q^2P^2 - P^2Q^2);$$

with (17), this gives

$$4A(qp) = -i(QPQP + iQP - P^2Q^2)$$

and by the further repetitive application of (17) then

$$4A(qp) = -i(2i(QP + PQ)).$$

iii) We have

$$A(q^3) = Q^3. \tag{18}$$

Setting $A(q^3) =: X$, it immediately follows from $\{q^3, q\} = 0$ that

$$[X, Q] = 0,$$

and, from $\{q^3, p\} = 3q^2$ with $(*)$ and (13),

$$[X, P] = 3iQ^2.$$

Trivially, in an associative algebra, we have

$$[Q^3, Q] = 0,$$

and because of (17)

$$[Q^3, P] = Q^3P - PQ^3 = Q^2PQ - PQ^3 + iQ^2 = \ldots = 3iQ^2.$$

From here, it immediately follows that $Y = X - Q^3$ commutes with P and Q, and then, because

$$[Y, PQ] = YPQ - PQY = [Y, P]Q + P[Y, Q] = 0,$$

also with PQ, and, by entirely analogous reasoning, also with QP.

This can be used in the following manner. Because $\{q^3, qp\} = 3q^3$ we get from (16) and $(*)$ that

$$[X, (1/2)(QP + PQ)] = 3iA(q^3) = 3iX,$$

but since $Y = X - Q^3$ commutes with QP and PQ, we also get

$$\begin{aligned} [X, (1/2)(QP + PQ)] &= [Q^3, (1/2)(QP + PQ)] \\ &= (1/2)(Q^4P + Q^3PQ - QPQ^3 - PQ^4), \end{aligned}$$

and, after further repeated applications of (17),

$$[X, (1/2)(QP + PQ)] = -3iQ^3;$$

thus $X = Q^3$.

iv) With an analogous calculation, it can be shown that

$$A(p^3) = P^3. \tag{19}$$

v) We have

$$A(q^2p) = (1/2)(Q^2P + PQ^2). \tag{20}$$

From $\{q^3, p^2\} = 6q^2p$ it follows from $(*)$, (18) and (17) that

$$\begin{aligned} 6iA(q^2p) &= [Q^3, P^2] = Q^3P^2 - P^2Q^3 = Q^2PQP + iQ^2P - P^2Q^3 \\ &= \ldots = 3i(Q^2P + PQ^2). \end{aligned}$$

vi) And again by the same methods, we arrive at

$$A(qp^2) = (1/2)(QP^2 + P^2Q). \tag{21}$$

vii) On the basis of the formulas (18) and (19), we have

$$(1/9)A\{q^3, p^3\} = (-i/9)[Q^3, P^3].$$

From this, we further get, after a few similar calculations (or, when we don't care about the purity of our method, quickly by returning to (12)),

$$(1/9)A\{q^3, p^3\} = -2/3 - 2iQP + Q^2P^2.$$

And then by (20) and (21), from what is in any case a tedious computation, we finally arrive at

$$\begin{aligned}(1/3)A\{q^2p, p^2q\} &= (-i/12)[(Q^2P + PQ^2), (P^2Q + QP^2)]\\ &= -1/3 - 2iQP + Q^2P^2.\end{aligned}$$

From (14) and (15) we have the two equations giving the contradiction promised in part i) of the proof. □

EXERCISE 5.12. Supply the missing calculations in the above proof.

Remark 5.13. The operators used here operate on a dense subspace of $L^2(\mathbb{R})$ and came into being with the help of the irreducible representation π^m_{SW}. We can now attempt to extend the area of applicability of quantization, by enlarging the space on which the operators of the quantization operate. For instance, we can try to avoid giving the operators as arising from an irreducible representation. ABRAHAM–MARSDEN ([**AM**], pp. 435–439) demonstrate the impossibility of extending the quantization to $\mathfrak{p}$ for the case of the space of functions with values in a finite–dimensional vector space, in which the steps presented here are just slightly generalized.

5.5. Towards the general case

Kirillov sketches ([**Ki**], pp. 241 f.) the following picture of the process for converting from classical mechanics to quantum mechanics. Although this has already been described at various occasions, we reproduce it yet once more to provide a unifying view for the general problems we will now handle.

In *classical mechanics*, the physical quantities are real functions $f \in \mathcal{F}(M)$. The symplectic manifold M is, in the first view, the phase space, thus the cotangent bundle T^*Q to a configuration space Q; but in actuality the conversion does not only concern itself with this situation. The time course of a physical system is fixed by giving a *Hamiltonian vector field* $X_H \in V(M)$, where $H \in \mathcal{F}(M)$ is the *energy* of the system. Then the traversal of an integral curve γ of X_H describes the time course of a quantity $f \in \mathcal{F}(M)$ and is given by the generalized Hamiltonian equation

$$\dot{f} = \{f, H\}.$$

A group G is called a symmetry group of the given system, when it operates via symplectic transformations on M.

In *quantum mechanics* the phase space, or the symplectic manifold M, is replaced by the projective space $\mathbb{P}(\mathcal{H})$ to an appropriate Hilbert space $\mathcal{H}$ for the system. The physical quantities f are then carried over to self–adjoint operators $\widehat{f}$, which then operate on (or, more precisely, in) $\mathcal{H}$ (on the rather subtle problem of the appropriate space of definition of these operators, we have no room here to elucidate). The *value* of the quantity f or $\widehat{f}$ that is the state of the system is then described by a unit vector $v_1 \in \mathcal{H}$ which is a random variable with the distribution function $p(s) = (E_s v_1, v_1)$, where E_s is the spectral projection measure of the operator $\widehat{f}$. This means that in order to describe the state of the system by v_1, the quantities f must have a fixed value a, for which v_1 is an eigenvector with eigenvalue a.

The time course is then given by a one parameter group of unitary operators on $\mathcal{H}$ of the form

$$U(t) = e^{\frac{ih}{2\pi} t\hat{H}},$$

where h is Planck's constant. $\hat{H}$ is here a self–adjoint operator, called the *energy operator*, so that then the time course of a quantity $\widehat{f} = F$ satisfies

$$\dot{F} = \frac{ih}{2\pi}[\hat{H}, F].$$

G is a symmetry group of the system, when G operates as unitary operators on $\mathcal{H}$.

A *quantization* is now the process of constructing, for a given classical system, a *corresponding* quantum system. It is however very difficult (when it can be done at all) to give a unique interpretation to the word *corresponding*. Classical mechanics should, in all possible interpretations, be understandable as the limit case of quantum mechanics for $h \to 0$. It should not be expected that the reversal of this process is *per se* unique. Be that as it may, Kirillov showed that this process is, in many known and simple cases, independent of the choice of the quantization process. In common with the sources we here cite Kirillov ([**Ki**], p. 243), Abraham–Marsden ([**AM**], pp. 433 f.), and Guillemin–Sternberg ([**GS**], p. 265), and, in view of the experience gained in Sections 5.1 through 5.4, we can attempt the following scheme.

We will consider in $\mathcal{F}(M)$ the physical quantities of a given physical system and select from them certain ones, the so-called *primary quantities*

f, which relative to the Poisson bracket generate a Lie algebra $\mathfrak{p}$. The transformation to a quantum system should now proceed so that

$$f \mapsto \widehat{f}$$

is an $\mathbb{R}$–linear map of $\mathfrak{p}$ into a Lie algebra of (self–adjoint) operators on a Hilbert space $\mathcal{H}$ with

$$(*) \qquad \widehat{\{f_1, f_2\}} = c\,[\widehat{f_1}, \widehat{f_2}] \quad \text{for } f_1, f_2 \in \mathfrak{p} \subset \mathcal{F}(M).$$

In Kirillov [**Ki**], we have $c = \frac{ih}{2\pi}$, where, on the other hand, the Poisson bracket has the opposite sign from the one we have here used. In ABRAHAM–MARSDEN [**AM**] as well as in GUILLEMIN–STERNBERG [**GS**], c is set equal to $-i$. Further, it is required that the constant functions are contained in $\mathfrak{p}$ and that we then have

$$(**) \qquad \widehat{1} = 1 = \mathrm{id}_{\mathcal{H}}.$$

A rule which delivers exactly this is called a *prequantization.* Otherwise (see, for example, ABRAHAM–MARSDEN ([**AM**], p. 434)) for a so-called *full quantization* the operation of the $\hat{f}$'s must satisfy an irreducibility condition in the sense that there be no proper general invariant subspace, or at least, only finitely many. For $\mathfrak{p} = \mathcal{F}_{\leq 2}$, the construction in Sections 5.1–5.3 gave a full quantization, since it arose as the infinitesimal representation of an irreducible representation π^m_{SW} of the Jacobi group. The statement of the Groenewold–van Hove Theorem 5.11 says that the domain of this quantization cannot be extended.

We can now follow a somewhat different line, which is here taken from KIRILLOV ([**Ki**], p. 243). And this begins by taking, for $M = T^*\mathbb{R}^n$, the subalgebra $\mathfrak{p}$ of $\mathcal{F}(M)$ spanned by the linear functions in $p_1, \ldots, p_n$, and the commutative algebra $\mathfrak{p}_0$ of functions in $\mathfrak{p}$ which depend only on $q_1, \ldots, q_n$. This algebra is infinite–dimensional and has the property that

$$\{f, g\} \subset \mathfrak{p}_0 \quad \text{for all } f \in \mathfrak{p},\ g \in \mathfrak{p}_0.$$

In agreement with the terminology of Kirillov, this algebra $\mathfrak{p}$ will be called the algebra of *primary quantities.* Under known conditions, it gives rise to a representation of the Poisson algebra $\mathcal{F}(M)$, which then, on restriction to $\mathfrak{p}$, satisfies a minimality condition for the representation space $\mathcal{H}$, which, it turns out, corresponds to the above–mentioned condition for full quantization.

The construction of a representation of the Lie algebra of primary quantities $\mathfrak{p}$.

i) The starting point for this construction is the map

$$\mathcal{F}(M) \ni f \mapsto X_f \in \mathrm{Ham}(M),$$

which gives the vector field X_f, after Theorem B.4, as a derivation on $\mathcal{F}(M)$, and therefore as a linear differential operator. This extends to

$$f \mapsto \widehat{f} = -iX_f$$

and satisfies, because $X_{\{f,g\}} = -[X_f, X_g]$ (see Corollary 3.23), the condition $(*)$ with $c = -i$. Since the image of the constant functions in $\mathrm{Ham}(M)$ is zero, this procedure must be altered if we wish $(**)$ to be satisfied. Therefore Kirillov takes, with the 1–form α on M,

$$(+) \qquad \widehat{f} := -iX_f + f + \alpha(X_f).$$

In this way, $(*)$ is satisfied so long as for α we have $d\alpha = \omega$.

EXERCISE 5.14. Using the relation 1) from Section 3.1, verify that

$$d\alpha(X_1, X_2) = L_{X_1}(\alpha(X_2)) - L_{X_2}(\alpha(X_1)) - \alpha\big([X_1, X_2]\big).$$

In the special case of $M = T^*Q$ one can take $-\alpha = \vartheta = \sum p_i dq_i$, and then $(+)$ has the form

$$\widehat{f} = -i\sum_{j=1}^{n}\left(\frac{\partial f}{\partial p_j}\frac{\partial}{\partial q_j} - \frac{\partial f}{\partial q_j}\frac{\partial}{\partial p_j}\right) + f - \sum_{j=1}^{n} p_j\frac{\partial f}{\partial p_j},$$

and so in particular

$$\widehat{p}_j = -i\frac{\partial}{\partial q_j}, \qquad \widehat{q}_j = i\frac{\partial}{\partial p_j} + q_j.$$

ii) If ω is not exact, one can use the above construction in a generalized form. In this case, by using the fact that ω is closed and thus locally exact, a cocycle h (see Section A.2) is constructed which has an associated vector bundle E of rank 1, thus a *line bundle.* Then the construction is generalized by assigning sections $\widehat{f}$ of this bundle to the primary quantities f as in $(+)$. This then goes through (see Theorem 1 of KIRILLOV ([**Ki**], p. 245)), in the case that the appropriate normalization of the integral of ω

$$\int_\gamma \omega$$

over an arbitrary 2–cycle γ in M is a whole number (an integral multiple of Planck's constant h in the normalization of Kirillov). It is further shown by Kirillov that

THEOREM 5.15. *There exists a one-to-one correspondence between the equivalence classes of representations of* $\mathfrak{p}$ *and the elements of the cohomology group* $H^1(M, \mathbb{C}^*)$.

In particular for connected M, $H^1(M, \mathbb{C}^*)$ is trivial, and therefore this quantization is unique.

In the general picture constructed so far, nothing has been said about the Hilbert space $\mathcal{H}$, nor has it been required that the operators $\widehat{f}$ should operate as self–adjoint operators, or that the choice of f is required to be restricted to $\mathfrak{p}$. The operators which appear in (+) are (differential) operators which operate on $\mathcal{F}(M)$ or on the space $\Gamma(E)$ of sections of the line bundle E. As already mentioned in the discussion of the quantization concept, it is desirable that the space on which the operators $\widehat{f}$ corresponding to f operate be as small as possible (so that the operation is irreducible or at least of finite multiplicity). In the explicit discussion of the examples $M = T^*Q \simeq \mathbb{R}^{2n}$ in Sections 5.1–5.3, there appears, somewhat informally, the space $\mathcal{H} = L^2(\mathbb{R}^n)$, which then is the desired space; namely, for $n = 1$ we get the quantization of $\mathfrak{g}^J$ with the help of the Schrödinger–Weil representation π_{SW} of the Jacobi group $G^J(\mathbb{R})$. In the next most general case $M = T^*Q$ we see that $\mathcal{H} = L^2(Q, d\mu)$, with an appropriate measure $d\mu$, is the proper space. Now for the most general situation we will greatly simplify the discussion, but still hope to make it clear that the choice of $\mathfrak{p}$ makes it possible to manage with this *small* Hilbert space as the domain for the operators $\widehat{f}$.

When M is not the cotangent space to a manifold, one would expect a Hilbert space of functions in $n = (1/2)\dim M$ variables. The local separation of variables in q and p via the form

$$\omega = \sum dq_j \wedge dp_j$$

is perhaps not globally possible. Halving the dimension is however possible with the help of the local concepts. This is given by defining on M a *Lagrange distribution* L; that is, by defining at every $m \in M$, a Lagrangian subspace L_m in T_mM (on which ω_m vanishes), so that these spaces *vary from point to point in a differentiable way.* (This can, like the concept of differentiable vector spaces, be made more precise). Analytically this leads to the choice of a maximal commutative subalgebra $\mathcal{F}_0(U) \subset \mathcal{F}(U)$, relative to $\{\,,\,\}$, for all neighborhoods $U \subset M$. Then we have

$$\mathcal{F}_0(U) = \{f \in \mathcal{F}(U) | X_f = 0 \text{ for all } X \in L\},$$

and $\Gamma_0(E, U)$ denotes the space of sections of $\Gamma(E)$ for which the functions $f \in \mathcal{F}_0(U)$ are assigned operators $\widehat{f}$, in accordance with the rule in ii), that operate as multiplication operators. This leads to a subspace $\Gamma_0(E)$ of the space of sections of $\Gamma(E)$, which can also be characterized as the space of sections which are annihilated by covariant differentiation (see Section A.4) along arbitrary $X \in L$. The primary quantities $f \in \mathfrak{p}$ are excellent, because they satisfy

$$\{f, \mathcal{F}_0(U)\} \subset \mathcal{F}_0(U) \quad \text{for all } U.$$

This condition implies that the operators $\widehat{f}$ associated to f leave $\Gamma_0(E)$ stable. In the special case of $M = T^*\mathbb{R}^n$ this translates to the result that the f are primary, that is, sums of arbitrary functions of $q_1, \ldots, q_n$ and linear functions of $p_1, \ldots, p_n$. $\mathcal{H}$ is then the completion of $\Gamma_0(E)$ relative to a fitting scalar product, so that the operators are selfadjoint.

Remark 5.16. This last briefly mentioned theme is introduced in KIRILLOV ([**Ki**], p. 241–248) with somewhat more, but still not full, detail, and shows that at least the topics introduced in the previous chapters and in the appendices can be collected into a general theory. The general theory goes yet further, but requires yet more application of functional analysis.

I hope that the interested reader is now sufficiently prepared to study the original sources: ABRAHAM-MARSDEN [**AM**], GUILLEMIN–STERNBERG [**GS**], KIRILLOV [**Ki**], WALLACH [**Wa**] and WOODHOUSE [**Wo**]. In particular, we refer the reader one last time to WALLACH [**Wa**], which has an appendix by R. HERMANN that is a readable introduction to quantum mechanics with many historical and critical remarks.

¿Te cojí? Yo no sé
si te cojí, pluma suavísima,
o si cojí tu sombra.

J.R. JIMÉNEZ

Differentiable Manifolds and Vector Bundles

The following definitions are standard in function theory and differential geometry, and also pass into symplectic geometry. The reader will find this material covered in [**AM**], p. 31 ff., or [**A**], pp. 77 ff. and 163 ff. The following presentation is also influenced by the books of H. Cartan: *Formes Differentielles* [**C**], Hollman and Rummler: *Alternierende Differentialformen* [**HR**], Sternberg: *Lectures on Differential Geometry* [**St**], and Chern: *Complex Manifolds without Potential Theory* [**Ch**], as well as the article by Kähler: *Der innere Differentialkalkül* [**K2**].

A.1. Differentiable manifolds and their tangent spaces

The concept of differentiable manifold is fundamental to the study of symplectic (as well as Riemannian) geometry, and so we begin with the definition of a p-dimensional ($p \in \mathbb{N}$) differentiable manifold. Here, differentiable will mean infinitely differentiable, although in many cases this will be stronger than needed.

Definition A.1. A connected Hausdorff topological space M with a countable basis is called a *p-manifold* if every point $m \in M$ lies in an open neighborhood $U \subset M$ homeomorphic to an open neighborhood of $\mathbb{R}^p$.

In particular, M is locally compact, and is also referred to as a *locally Euclidian space of dimension* p.

DEFINITION A.2. A *chart* for M is a pair (φ, U) (also referred to simply as φ), with U a neighborhood in M and $\varphi : U \to \mathbb{R}^p$ a homeomorphism onto an open subset $\varphi(U)$ of $\mathbb{R}^p$.

U will be called a *coordinate neighborhood*, φ the *coordinate map*, and φ_i (the ith component of φ) the ith *coordinate function*. The coordinates in $\mathbb{R}^p$ of the image of U under φ will be given by the coordinates $x = (x_1, \ldots, x_p)$. So for $m \in U$, $x = \varphi(m)$, we have $x_i = \varphi_i(m)$, $i = 1, \ldots, p$.

DEFINITION A.3. Two charts $\varphi : U \to \mathbb{R}^p$ and $\varphi' : U' \to \mathbb{R}^p$ are called *compatible* when the *transformation* functions

$$\varphi' \circ \varphi^{-1}\big|_{\varphi(U \cap U')} \quad \text{and} \quad \varphi \circ \varphi'^{-1}\Big|_{\varphi'(U \cap U')}$$

are differentiable.

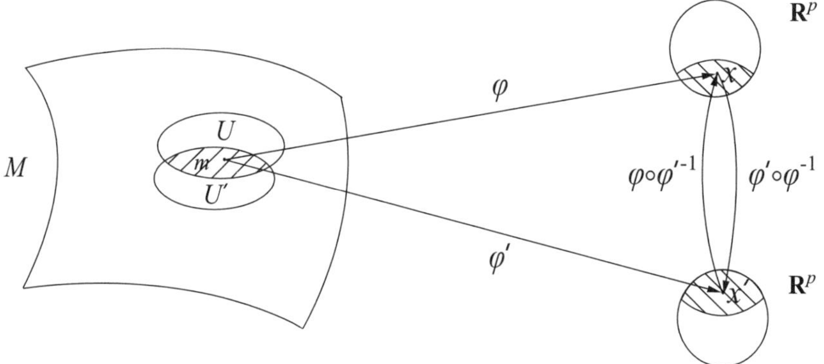

In general, we will abbreviate this to *M is a differentiable manifold.*

Analogous to this concept of real manifold is the concept of a *complex differentiable manifold* of dimension p. In this case each chart φ gives a homeomorphism of an open neighborhood $U \subset M$ to an open set of $\mathbb{C}^p$. One then requires that the transformation function between any two charts (φ, U) and (φ', U') be a holomorphic function.

There is not enough space here to cover the many examples that can and do arise in the literature. For this reason only those examples which show up frequently in this text will be discussed.

i) Open submanifolds. Let M be a differentiable manifold with atlas α, and M_0 an open set in M. Then the family $(M_0 \cap U, \varphi|_{M_0 \cap U})$, with (φ, U) in α, is an atlas α_0 on M_0. Such M_0, with the given differential structure, is called an open submanifold.

ii) p-dimensional submanifolds of $\mathbb{R}^p$. In FORSTER ([**F**], p. 128) and LANG ([**L5**], p. 363), one finds the following definition.

DEFINITION A.4. A subset $M \subset \mathbb{R}^q$ is called a *p-dimensional differentiable submanifold of* $\mathbb{R}^q$ if for every point $m \in M$ there exist an open neighborhood $V \subset \mathbb{R}^q$ and differentiable functions $f_i : V \to \mathbb{R}$ $(i = 1, \ldots, q-p)$ with

(1) $M \cap V = \{x \in V : f_1(x) = \ldots = f_{q-p}(x) = 0\}$,

(2) $\operatorname{rank} Df(m) = q - p$,

where

$$Df = \frac{\partial(f_1, \ldots, f_{q-p})}{\partial(x_1, \ldots, x_q)} = \begin{pmatrix} \frac{\partial f_1}{\partial x_1} & \cdots & \frac{\partial f_1}{\partial x_q} \\ \vdots & & \vdots \\ \frac{\partial f_{q-p}}{\partial x_1} & \cdots & \frac{\partial f_{q-p}}{\partial x_q} \end{pmatrix}$$

is the Jacobian of $f = (f_1, \ldots, f_{q-p})$.

It can be shown that such a differentiable submanifold is a differentiable manifold in the previous sense. Among the simplest examples are the *hypersurfaces*, with $q = p + 1$. Important examples of this class are the *hyperplanes*

$$H^p = \{x \in \mathbb{R}^{p+1} : f(x) = 0\},$$

where f is a linear function

$$f(x) = a_0 + \sum_{i=1}^{p+1} a_i x_i, \quad a_0, \ldots, a_{p+1} \in \mathbb{R}, \ \sum_{i=1}^{p+1} {a_i}^2 \neq 0,$$

as well as the *p-sphere*

$$S^p = \left\{ x \in \mathbb{R}^{p+1} : \sum_{i=1}^{p+1} {x_i}^2 = 1 \right\}.$$

iii) Complex projective space $\mathbb{P}^p(\mathbb{C})$. This is

$$\mathbb{P}^p(\mathbb{C}) = \{(x)_\sim : z \in \mathbb{C}^{p+1} \backslash \{0\}\},$$

the space of all complex lines in $\mathbb{C}^{p+1}$ passing through the origin; that is, $z \sim z'$ if there is a $\lambda \in \mathbb{C}^*$ with $z_i' = \lambda z_i$ for $i = 1, \ldots, p+1$. In this case, an atlas α, for $i = 1, \ldots, p+1$, is given by

$$\begin{aligned} \varphi^{(i)} : U_i = \{(z)_\sim : z_i \neq 0\} &\to \mathbb{C}^p, \\ m = (z)_\sim &\mapsto x = \left(\frac{z_1}{z_i}, \ldots, \frac{\widehat{z_i}}{z_i}, \ldots, \frac{z_{p+1}}{z_i} \right) \end{aligned}$$

(where the symbol $\widehat{\ }$ indicates that this entry should be omitted). Clearly the transformation functions are holomorphic; for instance, $\varphi = \varphi^{(1)}$ and

$\varphi' = \varphi^{(p+1)}$. We have

$$\begin{aligned}\varphi(m) &= x = \left(\frac{z_2}{z_1}, \dots, \frac{z_{p+1}}{z_1}\right), \\ \varphi'(m) &= x' = \left(\frac{z_1}{z_{p+1}}, \dots, \frac{z_p}{z_{p+1}}\right);\end{aligned}$$

it follows that

$$\varphi' \circ \varphi^{-1}\big|_{\varphi(U_1 \cap U_{p+1})}(x) = \left(\frac{1}{x_p}, \frac{x_1}{x_p}, \dots, \frac{x_{p-1}}{x_p}\right)$$

is then holomorphic, since $x_p = z_{p+1}/z_1 \neq 0$ in $U_1 \cap U_{p+1}$.

Morphisms of differentiable manifolds. The appropriate concept of maps for differentiable manifolds are those maps which are compatible with the differential structures.

DEFINITION A.5. Let M and N be differentiable manifolds of dimensions p and q, respectively. A map $F : M \to N$ is called a *differentiable map* or a *morphism* precisely when the following are satisfied:

i) F is continuous, and

ii) F is given locally by differentiable maps.

This means that there exist atlases α on M and β on N which fix the underlying differential structures of the manifolds, so that, for charts (φ, U) from α and (ψ, V) from β, on $W := U \cap F^{-1}(V)$ the composition

$$\varphi(W) \underset{\varphi^{-1}}{\longrightarrow} W \underset{F|_W}{\longrightarrow} V \underset{\psi}{\longrightarrow} \psi(V)$$

is differentiable.

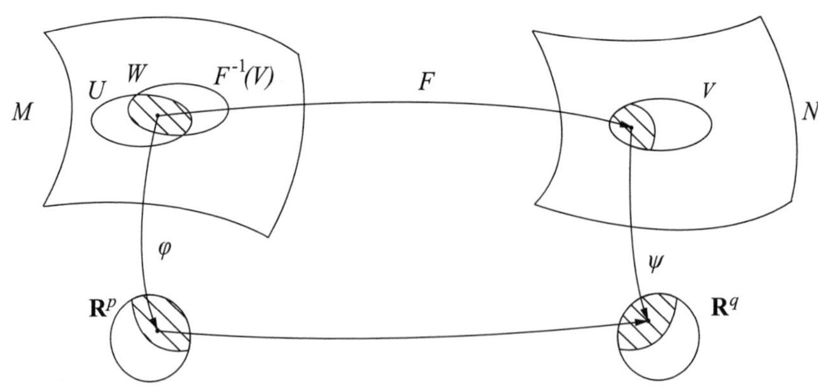

Remark A.6.

(1) The condition ii) is independent of the choice of atlases for M and N.

(2) The phrase F *is given by differentiable maps* should have the meaning that $\psi \circ F \circ \varphi^{-1}$ gives a q–tuple of differentiable functions in p variables in the local coordinates.

(3) The composition of differentiable maps is again a differentiable map.

DEFINITION A.7. A morphism of differentiable manifolds $F : M \to N$ is called an *isomorphism* or a *diffeomorphism*, when F is a homeomorphism to N and $F^{-1} : N \to M$ is also a morphism.

As an indication of just how general the concept of a submanifold of $\mathbb{R}^q$ is, we have the following statement.

THEOREM A.8. (Whitney) *A differentiable manifold of dimension p is diffeomorphic to a submanifold of $\mathbb{R}^{2p+1}$.*

Differentiable functions. For $N = \mathbb{R}$, we get, as a special case of the definition of a differentiable map, the concept of a *differentiable function* on a differentiable manifold M. For an open subset V of M, denote by $\mathcal{F}(V)$ the ring of differentiable functions defined on V, namely those functions $f : V \to \mathbb{R}$ such that, for every chart (φ, U) of an atlas α of M with $U \cap V \neq \emptyset$, the function $f \circ \varphi^{-1}|_{\varphi(U\cap V)}$ can be given by a differentiable function in the p variables $x_1, \dots, x_p$.

The function $f \circ \varphi^{-1}$ for a *typical chart* (φ, U) will, in agreement with the practice found in the literature, also be denoted simply by f. Thus $f(m) = f(x)$ for the value of the function f for which $m \in V$ in the chart (φ, U) corresponds to the point with coordinates x.

Analogously, we denote by $\mathcal{O}(V)$ the ring of holomorphic functions for an open set V of a complex differentiable manifold.

Tangent spaces. There are many ways in which one may assign *the* tangent space to a point m of a differentiable manifold M. The central point is the demonstration that it is a vector space which has the same dimension as M. The following definition formalizes the visual image that the tangent space in $m \in M$ is the set of the tangent vectors to all curves on M which pass through the given point.

DEFINITION A.9. Let m be a point in M. Then the *tangent space* of M at m is

$$T_m M := Cu_m M/_\sim,$$

where Cu_mM is the class of all pairs (J, γ), where J is an open interval of $\mathbb{R}$ containing 0 and $\gamma : J \to M$ is a differentiable map with $\gamma(0) = m$, with an equivalence relation defined on Cu_mM by

$$(J, \gamma) \sim (J', \gamma')$$

precisely when

$$\frac{d(\varphi_i \circ \gamma)}{dt}(0) = \frac{d(\varphi_i \circ \gamma')}{dt}(0) \qquad (i = 1, \ldots, p)$$

for a chart (φ, U) with $m \in U$.

With the help of the chain rule, it can be shown that this definition is independent of the choice of the chart.

Remark A.10. When $M \subset \mathbb{R}^q$ is a submanifold, this definition simplifies as follows: $v \in \mathbb{R}^q$ is a tangent vector to M at $m \in M$ (thus $\in T_mM$) precisely when there is, for a given $\varepsilon > 0$, a map $\gamma : (-\varepsilon, \varepsilon) \to M$ with $\gamma(0) = m$ and $\dot{\gamma}(0) = v$.

This remark prepares the general statement that T_mM, in a natural way, has the structure of a vector space of dimension p.

THEOREM A.11. *For each chart (φ, U) with $m \in U$ there exists a bijection*

$$h : T_mM \to \mathbb{R}^p$$

given by

$$X_m := (J, \gamma)_\sim \longmapsto \left(\frac{d(\varphi_i \circ \gamma)}{dt}(0)\right)_{i=1,\ldots,p}.$$

Now, given a second chart (φ', U'), with $m \in U'$ and $\varphi'(m) = y$, and transformation functions $y = y(x) = \varphi' \circ \varphi^{-1}|_{\varphi(U \cap U')}(x)$, then we have for the map h' associated to φ', on account of the chain rule,

$$h'_i(X_m) = \sum_{j=1}^{p} \frac{\partial y_i}{\partial x_j}(x)\, h_j(X_m) \qquad (i = 1, \ldots, p).$$

It is thus clear that T_mM has a canonical vector space structure which is independent of the choice of coordinates. It is conventional, for a chart (φ, U) containing m with the coordinates x, to take as a basis for T_mM the dual members of the canonical basis e of $\mathbb{R}^p$, so

$$\left.\frac{\partial}{\partial x}\right|_m = \left(\left.\frac{\partial}{\partial x_1}\right|_m, \ldots, \left.\frac{\partial}{\partial x_p}\right|_m\right)$$

(the elements here can only be read as symbols). Then we write

$$T_mM \ni X_m = \sum_{i=1}^{p} a_i \frac{\partial}{\partial x_i}\Big|_m \quad \text{with } a_i \in \mathbb{R} \qquad (i = 1, \ldots, p).$$

The transformation to another chart (φ', U') with coordinates y then leads to the following calculation:

$$X_m = \sum_{j=1}^{p} b_j \frac{\partial}{\partial y_j}\Big|_m \quad \text{with } b_j = \sum_{i=1}^{p} \frac{\partial y_j}{\partial x_i}(x) a_i.$$

This interpretation is consistent because of how partial derivatives behave under transformation, and it makes clear the following alternative.

The interpretation of T_mM as the space of derivatives. Denote by $\mathcal{F}(m)$ the collection of all the differentiable functions locally defined at $m \in M$, thus all pairs (U, f) with U an open neighborhood of m of M and f a differentiable function on U. Further, denote by $\mathcal{F}_m$ the set of all equivalence classes $(U, f)_\sim =: \tilde{f}$ with

$$(U, f) \sim (U', f')$$

precisely when there is a neighborhood U'' of m contained in $U \cap U'$ with $f|_{U''} = f'|_{U''}$. Such an equivalence class $(U, f)_\sim$ is called a *germ of* f. $\mathcal{F}_m$ can be easily given the structure of a ring, and from the value $f(m)$ of f at m the germ $(U, f)_\sim$ also has a uniquely defined value $\tilde{f}(m)$ at m.

DEFINITION A.12. The *tangent space* of M at m is

$$T_mM = \mathrm{Der}\,(\mathcal{F}_m, \mathbb{R}).$$

Here $\mathrm{Der}\,(\mathcal{F}_m, \mathbb{R})$ is the set of all $\mathbb{R}$–linear maps

$$L : \mathcal{F}_m \to \mathbb{R}$$

with

$$L(\tilde{f} \cdot \tilde{g}) = \tilde{f}(m)L\tilde{g} + \tilde{g}(m)L\tilde{f} \quad \text{for all } \tilde{f}, \tilde{g} \in \mathcal{F}_m.$$

Such linear maps are called *derivations* and can be interpreted as directional derivatives on the basis of the following theorems, which show that Definitions A.9 and A.12 are equivalent. Note that, as the space of linear maps, the space of derivations has an $\mathbb{R}$ vector space structure in a natural way.

THEOREM A.13. *There is an isomorphism*

$$\psi : Cu_mM/_\sim \xrightarrow{\sim} \mathrm{Der}\,(\mathcal{F}_m, \mathbb{R})$$

given by

$$X_m = (J, \gamma)_\sim \longmapsto L_{X_m} \quad \textit{with} \quad L_{X_m}\tilde{f} = \frac{d(f \circ \gamma)}{dt}(0),$$

where f is a representative for the germ $\tilde{f} \in \mathcal{F}_m$.

If $\gamma^{(j)}$ is a curve with $\varphi_i \circ \gamma^{(j)}(t) = \delta_{ij} t$, we get the associated derivation, here denoted by L_j. Clearly

$$L_j \tilde{f} = \frac{d(f \circ \gamma^{(j)})}{dt}(0) = \frac{\partial (f \circ \varphi^{-1})}{\partial x_j}(x),$$

which, by a slight abuse of notation, may be also written as

$$L_j \tilde{f} = \frac{\partial}{\partial x_j}\Big|_m f.$$

And so this is the derivative of f in the direction of $\gamma^{(j)}$ (the x_j–axis in the chart (φ, U)).

Cotangent spaces. As background for the later introduction of differential forms, the following concept will be useful.

Definition A.14. Let $m \in M$. Then the *cotangent space* to the point m is

$$T_m^* M = \mathrm{Hom}\,(T_m M,\, \mathbb{R}),$$

i.e., the dual space of the tangent space, consisting of the $\mathbb{R}$–linear maps α_m from $T_m M$ to $\mathbb{R}$. These maps will also be called *cotangent vectors.*

We now introduce some conventional notation. Let $\alpha_m \in T_m^* M$ be the map

$$\begin{aligned} \alpha_m : T_m M &\longrightarrow \mathbb{R}, \\ X_m &\longmapsto \langle X_m, \alpha_m \rangle := \alpha_m(X_m). \end{aligned}$$

Since $T_m M$ is a p–dimensional $\mathbb{R}$ vector space, this is also true for $T_m^* M$. The standard basis $\frac{\partial}{\partial x}|_m$ of $T_m M$ associated to the chart φ with coordinates x corresponds to a dual basis, which will be denoted as $(dx)_m$. Thus $(dx)_m$ denotes the p–tuple of linear forms on $T_m M$ with

$$\Big\langle \frac{\partial}{\partial x_i}\Big|_m, (dx_j)_m \Big\rangle = \delta_{ij}.$$

An $\alpha_m \in T_m^* M$ with

$$\alpha_m = \sum a_i (dx_i)_m, \quad a_i \in \mathbb{R}\ (i = 1, \ldots, p)$$

would in the coordinates y of another chart φ' and the correspondingly constructed basis $(dy)_m$ be written as

$$\alpha_m = \sum b_i (dy_i)_m.$$

With the notation

$$x = x(y) = \varphi \circ {\varphi'}^{-1}(y)$$

for the transformation functions between these two charts, we can evaluate the coefficients:

$$b_i = \sum_{j=1}^{p} \frac{\partial x_j}{\partial y_i}(y)a_j.$$

Remark A.15. Every $f \in \mathcal{F}(m)$, and thus every differentiable function which can be defined on a neighborhood of m, gives rise to a cotangent vector $(df)_m$, namely the linear form defined by

$$\langle X_m, (df)_m \rangle = L_{X_m} f = \frac{d(f \circ \gamma)}{dt}(0)$$

(here γ again stands for a curve through m with the tangent vector X_m). This is in agreement with the notation dx_i in the case that $f = x_i$ is the ith coordinate function of the chart φ, and with the usual concept of the total differential. Since, relative to a basis $(dx)_m$, $(df)_m$ can be represented by

$$(df)_m = \sum a_i(dx_i)_m,$$

we have that $(dx)_m$ is the dual basis to $\frac{\partial}{\partial x}|_m$,

$$a_j = \left\langle \frac{\partial}{\partial x_j}\Big|_m, (df)_m \right\rangle = \frac{\partial f}{\partial x_j}(x) \text{ (more precisely, } a_j = \frac{\partial(f \circ \varphi^{-1})}{\partial x_j}(\varphi(m)).$$

Thus

$$(df)_m = \sum_{j=1}^{p} \frac{\partial f}{\partial x_j}(x)(dx_j)_m.$$

Maps of tangent and cotangent spaces. Let $F : M \to N$ be a morphism of differentiable manifolds, $m \in M$ and $F(m) = n \in N$. Then in the following way we get a map

$$(F_*)_m : T_m M \to T_n N$$

of the tangent spaces. Given $X_m \in T_m M$ represented by the curve (J, γ) in M and $\gamma(0) = m$; then $(J, F \circ \gamma)$ is a curve in N with $F(\gamma(0)) = n$ and its tangent vector can be taken as the image of X_m. Thus

$$(F_*)_m(X_m) = (J, F \circ \gamma)_\sim =: Y_n \in T_n N,$$

and this is well defined. Dual to this stands the map

$$F_m^* : T_n^* N \to T_m^* M$$

of the cotangent spaces given by the following procedure. For $\beta_n \in T_n^* N$ we take for the image $F_m^* \beta_n$ in $T_m^* M$ the form which for all $X_m \in F_m M$ is defined by

$$F_m^* \beta_n(X_m) := \beta_n((F_*)_m X_m).$$

Thus, in other words,

$$\langle X_m, F_m^* \beta_n \rangle = \langle (F_*)_m X_m, \beta_n \rangle.$$

F is called a *submersion* if F_{*m} is surjective for all $m \in M$.

A.2. Vector bundles and their sections

The concept, which is general in physics and particularly frequently encountered in symplectic geometry, of a vector field X on a manifold M, is easy to grasp. The concept consists of a map which assigns to each point a tangent vector $X_m \in T_mM$. Somewhat more delicate to state precisely is when such a vector field should be called continuous or differentiable. This succeeds quite easily, however, with the help of the concept of vector bundle. We will now study this concept and will follow the treatment by FORSTER in *Riemannsche Flächen* ([**FRF**], pp. 195 ff.), where one may find the proofs which are here omitted. As a more complete coverage of this material, ABRAHAM–MARSDEN ([**AM**], pp. 37 f.) and the early pages of CHERN [**Ch**] can be recommended.

Grossly stated, a *vector space bundle*, or, more frequently as well as more briefly, a *vector bundle*, is a manifold which assigns in a smooth way to each point m of a basis manifold M a copy of a standard vector space V. Here one may imagine, for instance, the collection of all m of $M = S^2$, the sphere in $\mathbb{R}^3$, with attached tangent spaces. This is then a 4–dimensional structure. We now make this both more general and more precise, and examine some of its properties.

We assume for now that $K = \mathbb{R}$ or $\mathbb{C}$.

DEFINITION A.16. Let E and M be topological spaces and $\pi : E \to M$ a continuous map. Assume further that each fiber $E_m = \pi^{-1}(m)$, $m \in M$, has the structure of an n–dimensional K vector space. $\pi : E \to M$ or more briefly just E is called a *K vector(space) bundle of rank n over M* precisely when the following condition is satisfied. For every point $m \in M$ there are an open neighborhood U and a homeomorphism h from $E_U := \pi^{-1}(U)$ to $U \times K^n$ with the following properties:

i) The projection map factors through h; that is, the following diagram is commutative:

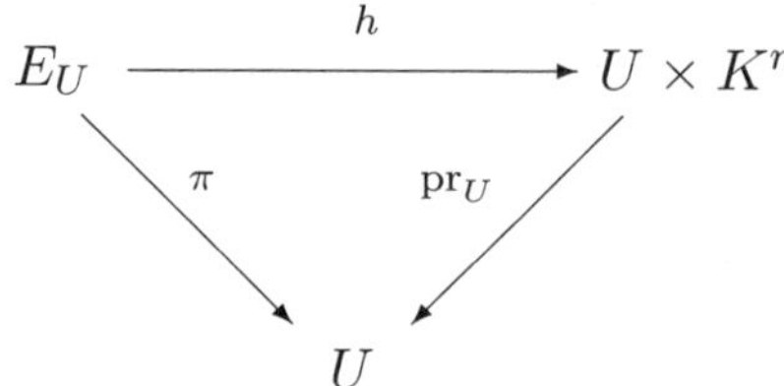

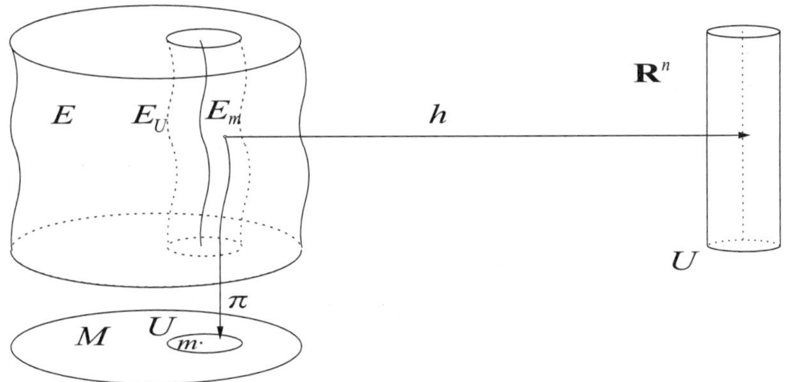

ii) For every $m \in U$, the map $h|E_m$ is a vector space isomorphism of E_m to $\{m\} \times K^n \cong K^n$.

The map $h : E_U \to U \times K^n$ is called a *linear chart* of E over U. If $\mathfrak{U} = (U_i)_{i\in I}$ is an open covering of M and the $h_i : E_{U_i} \to U_i \times K^n$ are linear charts, then the family $\mathfrak{A}$ of the h_i is called an *atlas of* E. A vector bundle of rank n is called *trivial* when there exists a global linear chart $h : E \to M \times K^n$.

Remark A.17. A vector bundle is thus defined to be locally trivial, and so, under local examination, the concept of a vector bundle delivers nothing new. It is only with a global study that they become interesting.

THEOREM A.18. *Let $E \to M$ be a vector bundle of rank n over M, and i an element from an index set I; let*

$$h_i : E_{U_i} \to U_i \times K^n, \qquad i \in I,$$

be the linear charts of an atlas of E. Then there are uniquely defined continuous maps

$$\psi_{ij} : U_i \cap U_j \to GL_n(K),$$

such that for the maps

$$h_{ij} := h_i \circ h_j^{-1} : (U_i \cap U_j) \times K^n \to (U_i \cap U_j) \times K^n$$

we have

$$h_{ij}(m, v) = (m, \psi_{ij}(m)v) \quad \textit{for all} \quad (m, v) \in (U_i \cap U_j) \times K^n.$$

On $U_i \cap U_j \cap U_k$ the cocycle relation *holds:*

$$\psi_{ij}\,\psi_{jk} = \psi_{ik}.$$

Remark A.19. As a matter of notation, the maps ψ_{ij} are called *transformation functions*, and the family (ψ_{ij}), $i, j \in I$ is called the atlas $\mathfrak{A} = (h_i)_{i \in I}$ associated to the *cocycle.*

DEFINITION A.20. Let M be a real or complex differentiable manifold, $E \to M$ an $\mathbb{R}$ or $\mathbb{C}$ vector space bundle of rank n over M, and

$$\mathfrak{A} = (h_i : E_{U_i} \to U_i \times K^n, i \in I)$$

an atlas of E. The atlas is called *differentiable* if the associated transformation functions ψ_{ij} are differentiable. Two differentiable atlases $\mathfrak{A}$ and $\mathfrak{A}'$ are called *compatible* if $\mathfrak{A} \cup \mathfrak{A}'$ is again a differentiable atlas.

It is easy to see that this compatibility is an equivalence relation. An equivalence class of compatible differentiable atlases is called a *differentiable linear structure* on E. A *differentiable vector bundle* is a vector bundle $E \to M$ equipped with a differentiable linear structure over a differentiable manifold. This definition deals with the real case. When V is a $\mathbb{C}$ vector space, everything that follows can be easily carried over.

Remark A.21. For a differentiable vector bundle on a differentiable manifold with the projection map $\pi : E \to M$, E is itself a differentiable manifold and π is a differentiable map.

A differentiable vector bundle $E \to M$ is called *differentiably trivial* if the differentiable linear structure contains an atlas which consists of but a single chart, $E \to M \times K^n$.

DEFINITION A.22. Let $E \xrightarrow{\pi} M$ be a differentiable vector bundle and $U \subset M$. A *differentiable section* of E over U is a differentiable map

$$f : U \to E \quad \text{with} \quad \pi \circ f = id_U.$$

The collection of all sections will be denoted by $\Gamma(E, U)$. This is again a K vector space.

Analogously one can, for every $r \in \mathbb{N}_0$, define the sections of class $\mathcal{C}^r$. The resulting spaces are then denoted by $\mathcal{C}^r_E(U)$.

A.3. The tangent and the cotangent bundles

Differentiable vector fields can now be defined as differentiable sections of the associated tangent bundle $E = TM$ of a differentiable manifold M. In this somewhat general framework, we will consider the *construction of vector bundles.*

For an open subset U of the differentiable manifold M, denote by

$$GL_n(\mathcal{F}(U))$$

the group of all invertible $n \times n$–matrices with coefficients in the space $\mathcal{F}(U)$ of differentiable functions on U. If $\mathfrak{U} = (U_i)_{i \in I}$ is an open covering of M, then denote by

$$Z^1(\mathfrak{U}, GL_n(\mathcal{F}))$$

the set of all 1–cocycles relative to $\mathfrak{U}$, that is to say, the families $(\psi_{ij})_{i,j \in I}$ with

$$\psi_{ij} \in GL_n(\mathcal{F}(U_i \cap U_j))$$

and

$$\psi_{ij}\psi_{jk} = \psi_{ik} \text{ over } U_i \cap U_j \cap U_k \text{ for all } i, j, k \in I.$$

If $\mathfrak{A}$ is a differentiable atlas of a vector bundle over M, then the family of the transformation functions of $\mathfrak{A}$ generate such a 1–cocycle. In the reverse direction, from every 1–cocycle from $Z^1(\mathfrak{U}, GL_n(\mathcal{F}))$ a differentiable vector bundle of rank n can be constructed.

THEOREM A.23. *Let M be a differentiable manifold, $\mathfrak{U} = (U_i)_{i \in I}$ an open covering of M and (ψ_{ij}) a family from $Z^1(\mathfrak{U}, GL_n(\mathcal{F}))$. Then there are a differentiable vector bundle $\pi : E \to M$ of rank n and a differentiable atlas*

$$(h_i : E_{U_i} \to U_i \times K^n)_{i \in I}$$

of E whose transformation functions are the given ψ_{ij}.

Of the proof, we will only say enough to show that the bundle $E \to M$ is given by the space

$$E' := \bigcup_{i \in I} U_i \times K^n \times \{i\} \subset M \times K^n \times I$$

by introducing an equivalence relation

$$(m, v, i) \sim (m', v', i')$$

precisely when

$$m = m' \quad \text{and} \quad v = \psi_{ii'}(m)v'.$$

Then, with the help of a few facts from topology, one can show that E arises as $E'/\sim$.

As an application we will now develop the concepts of the *tangent bundle* TM and the *cotangent bundle* T^*M. To this end, let M again be covered by $\mathfrak{U} = (U_i)_{i \in I}$ and, as we did earlier, denote by $\varphi^{(i)} : U_i \to K^p$ the coordinate map on the ith chart. Then, for the tangent bundle TM, we can take for the transformation functions ψ_{ij} those matrices which are the Jacobians of the transformations on the charts $\varphi^{(i)} \circ (\varphi^{(j)})^{-1}$; thus

$$\psi_{ij}(m) = J_{\varphi^{(i)} \circ (\varphi^{(j)})^{-1}}(\varphi^{(j)}(m)).$$

If $\varphi^{(i)}$ has the coordinates y and $\varphi^{(j)}$ the coordinates x, this says that from

$$\begin{array}{rcl} h_{ij} = h_i \circ h_j^{-1} : (U_i \cap U_j) \times K^p & \longrightarrow & (U_i \cap U_j) \times K^p, \\ (m,\, a) & \longmapsto & (m, \psi_{ij}(m)a = b), \end{array}$$

we may calculate that the vector $a = (a_\mu)$ in the x–coordinates has, in the y–coordinates, the components $b = (b_\nu)$ with

$$b_\nu = \sum_{\mu=1}^{p} \frac{\partial y_\nu}{\partial x_\mu}(\varphi^{(j)}(m))a_\mu.$$

For the *cotangent bundle* T^*M one correspondingly takes

$$\psi_{ij}(m) = {}^t J_{\varphi^{(j)} \circ (\varphi^{(i)})^{-1}}(\varphi^{(i)}(m)),$$

which for the vectors a^* and b^* is the transformation given by

$$b^*_\nu = \sum_{\mu=1}^{p} \frac{\partial x_\mu}{\partial y_\nu}(\varphi^{(i)}(m))a^*_\mu.$$

Remark A.24. As a set, these are

$$TM = \bigcup_{m \in M} T_m M, \text{ respectively, } T^*M = \bigcup_{m \in M} T^*_m M;$$

that is to say, they are the union of all tangent, respectively cotangent spaces, to all points $m \in M$. A differentiable vector field, thus a global differentiable section, $X : M \to TM$ can be written for a chart $(\varphi,\, U)$ with the coordinates $x = (x_1, \ldots, x_p)$ (with a slight but already familiar abuse of notation) by

$$X|_U = \sum_{\mu=1}^{p} a_\mu(x) \frac{\partial}{\partial x_\mu} \quad \text{with} \quad a_\mu \in \mathcal{F}(U).$$

For a chart (φ', U') with coordinates $y = (y_1, \ldots, y_p)$ we have, correspondingly,

$$X|_{U'} = \sum_{\nu=1}^{p} b_\nu(y) \frac{\partial}{\partial y_\nu} \quad \text{with} \quad b_\nu \in \mathcal{F}(U'),$$

and in the overlapping set $U \cap U'$ we have (so long as it's not empty) the transformation rule

$$b_\nu(y) = \sum_{\mu} \frac{\partial y_\nu}{\partial x_\mu}(y) a_\mu(x(y)) \quad \text{(contra)}.$$

Correspondingly, a field of cotangent vectors can be considered, which we call *differential forms of degree* 1 (or 1–*forms*), and can be introduced as differentiable global sections

$$\alpha : M \to T^*M$$

of the cotangent bundle. For these, we have analogously

$$\alpha|_U = \sum a^*_\mu(x)dx_\mu, \quad a^*_\mu \in \mathcal{F}(U),$$

respectively

$$\alpha|_{U'} = \sum b^*_\nu(y)dy_\nu, \quad b^*_\nu \in \mathcal{F}(U),$$

with the transformation rule

$$b^*_\nu(y) = \sum_\mu \frac{\partial x_\mu}{\partial y_\nu}(x(y))a^*_\mu(x(y)) \qquad \text{(co)}.$$

DEFINITION A.25. The differentiable global vector fields on M will be denoted by $V(M)$, and the 1–forms by $\Omega^1(M)$.

Remark A.26. In the older literature, particularly in many texts from physics, one finds the following definitions: A *contra*, respectively *covariant vector* (or one–fold *contra*, respectively *covariant tensor*) on M is a rule which assigns to a chart (φ, U) of M with coordinates x a system of p differentiable functions A_φ with the condition that for two charts (φ, U) and (φ', U') with $U \cap U' \neq \emptyset$ a relation as in (contra), respectively (co), exists between the function systems A_φ and $A_{\varphi'}$. In this sense the components of a vector field describe a one–fold contravariant tensor, and those from a degree 1 differential form a one–fold covariant tensor.

EXAMPLE A.27. For every differentiable function on M, its total differential df can be represented by the 1–form which in a chart (φ, U) with the coordinates x has the form

$$df = \sum_{\mu=1}^{p} \frac{\partial f}{\partial x_\mu} dx_\mu.$$

Maps of vector fields and 1–forms. The description of the maps of tangent and cotangent spaces at the end of Section A.1 leads immediately to the fact that vector fields and their dual objects are mapped by a differentiable map $F : M \to N$ in the following way. if F is *injective*, then to $X \in V(M)$ is assigned a vector field F_*X on $F(M) \subset N$ such that at every $n = F(m) \in F(M)$

$$(F_*X)_n = (F_*)_m X_m$$

is assigned. When $F(M)$ is a submanifold, one can prove that this vector field is again differentiable. And this is the case when F is an *embedding*; that is, all F_{*m} are injective and F is a homeomorphism of $F(M)$.

In order to pull back a 1–form $\beta \in \Omega^1(N)$ on M, the injectivity of F is not needed. $F^*\beta \in \Omega^1(M)$ is defined at every $m \in M$ so that

$$(F^*\beta)_m := F^*_m \beta_n \quad \text{with} \quad n = F(m).$$

This assignment gives a differentiable section of T^*M.

With a little more formalism it can be seen (see, for example, ABRAHAM–MARSDEN ([**AM**], p. 45)) that a differentiable map $F: M \to N$ gives rise to a differentiable map $TF : TM \to TN$ of the tangent bundles and analogously to a differentiable map $T^*F : T^*N \to T^*M$ of the cotangent bundles.

A.4. Tensors and differential forms

With the help of some *multilinear algebra* (see, for example GREUB [**Gb**], MARCUS [**Ma**], or, for the complete background, BOURBAKI ([**Bo**], Chapter III, §5)) one may carry this concept yet further. Given three (finite–dimensional) vector spaces T, T' and W, and a number $q \in \mathbb{N}$, let $L(T \times T', W)$ be the K vector space of bilinear maps

$$f : T \times T' \to W.$$

The tensor product of T and T', $T \otimes T'$, is then the (up to isomorphism, uniquely determined) K vector space with

$$L(T \times T', W) \simeq \mathrm{Hom}\,(T \otimes T', W) \text{ for all } W.$$

Let $L_q(T, W)$ be the K vector space of q–linear maps $f : T^q \to W$, and $A_q(T, W)$ the K vector space of all alternating q–linear maps $f : T^q \to W$. Then $\bigwedge^q T$ is the qth exterior power of T, thus the (up to isomorphism, uniquely determined) K vector space with

$$A_q(T, W) \simeq \mathrm{Hom}\,(\bigwedge^q T, W) \text{ for all } W.$$

Now put $L_q(T) := L_q(T, K)$ and $A_q(T) := A_q(T, K)$; so that, in particular,

$$L_1(T) = A_1(T) = \mathrm{Hom}\,(T, K) = T^*$$

is the dual space of T and $A_q(T) \simeq \bigwedge^q T^*$.

Let $t = (t_1, \ldots, t_n)$ be a basis of T and $t' = (t'_1, \ldots, t'_m)$ a basis of T'. Then

$$\begin{array}{lll} t \otimes t' = (t_i \otimes t'_j)_{i,j=1,\ldots,n} & \text{is a basis of} & T \otimes T', \\ t^q = (t_{i_1} \otimes \ldots \otimes t_{i_q})_{1 \le i_j \le n} & \text{is a basis of} & \bigotimes^q T, \\ \wedge t^q = (t_{i_1} \wedge \ldots \wedge t_{i_q})_{1 \le i_1 < \ldots < i_q \le n} & \text{is a basis of} & \bigwedge^q T, \end{array}$$

and, as we've already used,

$$t^* = (t_1^*, \ldots, t_n^*) \text{ with } t_i^*(t_j) = \langle t_i, t_j^* \rangle = \delta_{ij} \text{ is a basis of } T^*.$$

With these algebraic constructs we may now obtain yet more vector bundles E on M, whose fibers E_m are, for example, $\bigotimes^q T_m M$ or $\bigwedge^l T_m^* M$. In the generalization of the definitions of the previous section, an *l–fold co and q–fold contravariant tensor* on M is then, for example, a differentiable section

X of this bundle E such that X is a map which assigns to each $m \in M$ an element

$$X_m \in \bigotimes^l T_m^* M \otimes \left(\bigotimes^q T_m M \right),$$

and is differentiable in the following sense. For every point m from a chart φ of an atlas α of M, the components X_m relative to the standard bases associated to φ, given as described above by

$$t = \left. \frac{\partial}{\partial x} \right|_m \text{ for } T_m M, \text{ respectively, } t^* = (dx)_m \text{ for } T_m^* M,$$

are differentiable. Moreover a change of charts carries this assignment via an adequate assembly of transformation rules for both (co) and (contra).

Remark A.28. Without exploiting multilinear algebra, one can, as in Remark A.26, also give a definition of these in the style of the *old definition.* An l–fold co and q–fold contravariant tensor is a set of data which assigns to every chart (φ, U) of M a system

$$(X_\varphi)^{i_1 \ldots i_q}_{j_1 \ldots j_l} \qquad (1 \leq i_r, j_s \leq n;\ r = 1, \ldots, q,\ s = 1, \ldots, l)$$

of differentiable functions in $\varphi(U)$ with the condition that for two charts φ and φ' with $U \cap U' \neq \emptyset$ between the systems X_φ and $X_{\varphi'}$ a relation exists which gives for every index i a (contra) transformation formula and for every index j a (co) transformation formula.

Example A.29. As an example of a 1–fold contra and 2–fold covariant tensor, we have

$$(A_{\varphi'})^{i_1}_{j_1 j_2}(y) = \sum_{r_1, s_1, s_2} \frac{\partial y_{i_1}}{\partial x_{r_1}}(y) \frac{\partial x_{s_1}}{\partial y_{j_1}}(x(y)) \frac{\partial x_{s_2}}{\partial y_{j_1}}(x(y))(A_\varphi)^{r_1}_{s_1 s_2}(x).$$

Of greater significance, especially for physics, are those tensors which moreover realize a known symmetry condition. Before introducing these, we next consider an example of one of the fundamental objects of *Riemannian geometry.*

The Riemannian metric. For each $m \in M$, let a scalar product $\langle\ ,\ \rangle$ in $T_m M$ be given—that is, a symmetric positive definite bilinear form. In the canonical basis representation this means that for $v, w \in T_m M$

$$\langle v,\, w \rangle = \sum_{i,j} v_i w_j g_{ij}$$

with

$$g_{ij}(\varphi(m)) := \left\langle \left. \frac{\partial}{\partial x_i} \right|_m, \left. \frac{\partial}{\partial x_j} \right|_m \right\rangle \text{ for } i, j = 1, \ldots, p.$$

The assignment of an atlas α of M for each $m \in M$ and every chart (φ, U) allows one to assign to

$$m \longmapsto (g_{ij}(\varphi(m))) \quad \text{for} \quad m \in U$$

a system of functions, defining then a twofold covariant tensor. This tensor is called a *metric* or a *Riemannian fundamental tensor*, when

a) the matrix $(g_{ij}(x))$ for all $x \in \varphi(U)$ is symmetric and positive definite,

b) the functions $g_{ij}(x)$ are all differentiable, and

c) between the function systems of compatible charts, there is the following relationship between the transformation functions:

$$\tilde{g}_{ij}(y) = \sum_{r,s} \frac{\partial x_r}{\partial y_i}(x(y)) \frac{\partial x_s}{\partial y_j}(x(y)) g_{rs}(x(y)).$$

Naturally, a metric tensor can be better understood as the differentiable section of a bundle E over M which has as fibers E_m the *second symmetric power* $S^2T^*_mM$ of the cotangent space T^*_mM. It is also frequently fixed with the help of a *degree 2 symmetric differential form* in the following way (in the standard–chart (φ, U) with the coordinates x):

$$ds^2 = \sum g_{ij} dx_i \, dx_j.$$

A Riemannian fundamental tensor in this form gives rise to a Riemannian metric, which for $m, m' \in M$ and a curve γ linking the two points defines the *distance* between the points along γ by

$$\int_\gamma \sqrt{ds^2}.$$

With this the manifold receives yet another additional structure which is often of great help. One may derive, from a metric fundamental tensor, the Riemannian curvature tensor which is required for the formulation of the theory of general relativity. The following theorem is therefore of particular interest

THEOREM A.30. *On a differentiable manifold there is but one Riemannian fundamental tensor.*

Proof. See HOLMANN–RUMMLER ([**HR**], p. 108) or LANG ([**L2**], p.231). □

Of great significance is also the concept of *pseudo–Riemanniann manifold*, which comes from a slight weakening of the conditions on the metric fundamental tensor. In particular one weakens condition a) in the above definition to the requirement that the matrix $(g_{ij}(x))$ be symmetric and non–degenerate.

From total skew–symmetric covariant tensors one gets the (exterior) *qth degree differential forms*: differentiable sections $\alpha^{(q)}$ of the bundle E over M whose fibers E_m are the qth exterior powers $\bigwedge^q T_m^* M$ of the cotangent space are called exterior qth degree differential forms or simply q–forms. In the standard chart (φ, U) with the coordinates x, they have the *reduced representation*

$$\alpha^{(q)}\Big|_U = \sum_{1\le i_1<\ldots<i_q\le p} a_{i_1\ldots i_q}\, dx_{i_1}\wedge\ldots\wedge dx_{i_q}, \qquad a_{i_1,\ldots,i_q}\in\mathcal{F}(U).$$

Using the notation $I_q := \{(i_1,\ldots,i_q)|1\le i_1<\ldots<i_q\le p\}$, this is frequently more briefly written as

$$\alpha^{(q)} = \sum_{(i)\in I_q} a_{(i)} dx_{i_1}\wedge\ldots\wedge dx_{i_q} = \sum_{(i)\in I_q} a_{(i)} dx^q_{(i)}.$$

The set of exterior q–forms on M forms an $\mathbb{R}$ vector space, in which the value at the points $m\in M$ is declared to be the product. This space is denoted by $\Omega^q(M)$, and, putting $\Omega^0(M) := \mathcal{F}(M)$, we have the definition:

$$\bigoplus_{q=0}^{p} \Omega^q(M) =: \Omega(M).$$

The calculus of differential forms. The concept of differential forms allows one to define yet other operations in addition to taking sums. Here we will only give a short overview. Begin by letting $U\subset\mathbb{R}^p$ be an open set. It will later be made clear that all can be carried over to differential forms on manifolds.

1) As already noted, two differential forms

$$\alpha^{(q)} = \sum a_{(i)} dx^q_{(i)}, \quad \tilde{\alpha}^{(q)} = \sum \tilde{a}_{(i)} dx^q_{(i)} \in \Omega^q(U)$$

can be added:

$$\alpha^{(q)} + \tilde{\alpha}^{(q)} := \sum (a_{(i)} + \tilde{a}_{(i)}) dx^q_{(i)}.$$

Moreover, the multiplication of a differential form by a function $f\in\mathcal{F}(U)$ is well defined by

$$f\alpha^{(q)} := \sum f a_{(i)} dx^q_{(i)},$$

and therefore $\Omega^q(U)$ can be considered as an $\mathcal{F}(U)$–module.

Moreover, $\Omega(U) = \bigoplus \Omega^q(U)$ is itself a ring. The product is given by setting

$$(dx_{i_1} \wedge \ldots \wedge dx_{i_q}) \wedge (dx_{j_1} \wedge \ldots \wedge dx_{j_r}) := \chi(\pi) dx_{l_1} \wedge \ldots \wedge dx_{l_{q+r}},$$

where $\chi(\pi) = 0$ whenever all the i and j are not different, and otherwise $\chi(\pi)$ is the sign of the permutation π which carries $(i_1, \ldots, i_q, j_1, \ldots, j_r)$ to the *ordered form* $(l_1, \ldots, l_{q+r})$ with $1 \leq l_1 < \ldots < l_{q+r}$. Linearity then permits a product to be defined by

$$\begin{aligned} \Omega^q(U) \times \Omega^r(U) &\longrightarrow \Omega^{q+r}(U), \\ (\alpha^{(q)}, \tilde{\alpha}^{(r)}) &\longmapsto \alpha^{(q)} \wedge \tilde{\alpha}^{(r)}. \end{aligned}$$

This product is associative, but not commutative. It is called *exterior multiplication* and satisfies the commutation rule

$$\alpha^{(q)} \wedge \alpha^{(r)} = (-1)^{qr} \alpha^{(r)} \wedge \alpha^{(q)}.$$

From the interpretation of the differential forms $\alpha^{(q)}$ and $\alpha^{(r)}$ as alternating q–, respectively r–, linear forms on T_mM, we get that $\alpha^{(q)} \wedge \alpha^{(r)}$, through *antisymmetrization* of the operation of $\alpha^{(q)} \times \alpha^{(r)}$ on $(T_mM)^{q+r}$, defines a $(q+r)$–linear map.

Remark A.31. Differential forms may be added coefficient by coefficient and (by writing one after the other) may be associatively multiplied under the commutation rules

$$\begin{aligned} dx_i \wedge dx_j &= -dx_j \wedge dx_i \quad \text{for } i \neq j, \\ &= 0 \quad \text{otherwise.} \end{aligned}$$

For the specification of differential forms, one can exploit the already used ordered or reduced representation, so that an element of the *skew–symmetric representation* is usually written as

$$\alpha^{(q)} = \sum_{j_1, \ldots, j_q = 1}^{p} \frac{1}{q!} b_{j_1 \ldots j_q} \, dx_{j_1} \wedge \ldots \wedge dx_{j_q},$$

and so it then holds that

$$\begin{aligned} b_{j_1 \ldots j_q} &= 0, && \text{when } j_1, \ldots, j_q \text{ are not } q \text{ different integers,} \\ &= \chi(\pi) a_{i_1 \ldots i_q}, && \text{when } j_1, \ldots, j_q \text{ are pairwise different and } (i_1, \ldots, i_q) \text{ is carried by the permutation } \pi \text{ to } (j_1, \ldots, j_q). \end{aligned}$$

2) The assignment discussed at the end of Section A.3, which assigns to every $f \in \mathcal{F}(U) = \Omega^0(U)$ a 1–form $df \in \Omega^1(U)$, can be extended to an arbitrary q–form, and then delivers the *exterior differentiation*.

DEFINITION A.32. For a differential form $\alpha^{(q)}$ in the ordered representation

$$\alpha^{(q)} = \sum a_{(i)} dx_{i_1} \wedge \ldots \wedge dx_{i_q},$$

the *total differential* is defined to be

$$d\alpha^{(q)} := \sum da_{(i)} \wedge dx_{i_1} \wedge \ldots \wedge dx_{iq}, \qquad da_{(i)} := \sum_{k=1}^{p} \frac{\partial a_{(i)}}{\partial x_k} dx_k.$$

For further calculations $d\alpha^{(q)}$ is brought to the ordered or the skew–symmetric forms. The operation d is then extended linearly to the whole of $\Omega(U)$.

DEFINITION A.33. A differential form $\alpha \in \Omega(U)$ is called *closed* when $d\omega = 0$, and *exact* when there is a $\beta \in \Omega(U)$ with $\alpha = d\beta$.

The total differential d satisfies the following easily verified calculation rules for $\alpha \in \Omega^q(U)$ and $\beta \in \Omega^r(U)$:

$$\begin{aligned} d(\alpha + \beta) &= d\alpha + d\beta \quad (\text{for } q = r), \\ d(\alpha \wedge \beta) &= d\alpha \wedge \beta + (-1)^q \alpha \wedge d\beta, \\ d(d\alpha) &= 0. \end{aligned}$$

Poincaré's lemma says that for convex U every closed differential is exact. With the help of cohomology groups (see Section B.3) one may significantly sharpen this important statement.

3) In the interpretation of the differential as a differentiable section of a bundle, the appropriate behavior under transformation from one chart to another is already built in. To make our notation compatible, we postulate the following general behavior under *substitution of variables.* Let $U \subset \mathbb{R}^p$ be open with coordinates $x = (x_1, \ldots, x_p)$, $V \subset \mathbb{R}^r$ open with coordinates $y = (y_1, \ldots, y_r)$, and

$$\begin{aligned} F : U &\longrightarrow V, \\ x &\longmapsto F(x) = y, \end{aligned}$$

a differentiable map with the components $f_i(x_1, \ldots x_p)$, $i = 1, \ldots, r$. Every differential

$$\beta = \sum_{q=0}^{r} \sum_{(i)} b_{(i)} dy_{i_1} \wedge \ldots \wedge dy_{i_q} \in \Omega(V)$$

on V then assigns a differential on U by

$$F^*\beta = \beta F = \sum_{q=0}^{r} \sum_{(i)} b_{(i)} \circ F df_{i_1} \wedge \ldots \wedge df_{i_q}.$$

It is not hard to prove that this. In the special case of the earlier coordinate change with the given preconditions, the following holds.

Remark A.34. The map $F : U \to V$ induces a map

$$\begin{aligned} F^* : \Omega(V) &\longrightarrow \Omega(U), \\ \beta &\longmapsto F^*\beta = \beta F, \end{aligned}$$

which obeys the rules

$$\begin{aligned} (\beta + \beta')F &= \beta F + \beta' F, \\ (\beta \wedge \beta')F &= \beta F \wedge \beta' F, \end{aligned}$$

and therefore is a ring homomorphism. Since substitution and differentiation commute, we have

$$d(F^*\beta) = d(\beta F) = (d\beta)F = F^*(d\beta) \quad \text{for all} \quad \beta \in \Omega(V).$$

With this remark, it is clear that all the earlier operations defined only on $U \subset \mathbb{R}^p$, thus locally defined, are also globally defined, and so hold for differential forms on manifolds.

When a vector field $X \in V(M)$ is given, we may define yet another pair of operations on the differential forms $\Omega(M)$ on M.

4) The *inner multiplication* $i(X)$ decreases the degree of a differential form $\alpha^{(q)}$ of $\Omega^q(M)$ by one. It is defined by assigning to $i(X)\alpha^{(q)}$, for $X \in V(M)$, the map

$$i(X)\alpha^{(q)}(X_1, \dots, X_{q-1}) := \alpha^{(q)}(X, X_1, \dots, X_{q-1}).$$

It is not hard to see (see ABRAHAM–MARSDEN ([**AM**], p. 115)) that $i(X)\alpha^{(q)} \in \Omega^{q-1}(M)$ really holds, and that $i(X)$ satisfies the following rules:

i) $i(X)$ is $\mathbb{R}$–linear with

$$i(X)(\alpha^{(q)} \wedge \beta^{(r)}) = i(X)\alpha^{(q)} \wedge \beta^{(r)} + (-1)^q \alpha^{(q)} \wedge i(X)\beta^{(r)},$$

ii) $i(fX)\alpha^{(q)} = f i(X)\alpha^{(q)}$,

iii) $i(X)df = L_X f$ for $f \in \mathcal{F}(M)$, and

iv) $F^*(i(F_*X)\beta^{(q)}) = i(X)F^*\beta^{(q)}$ for $F : M \to M'$ a diffeomorphism and $\beta^{(q)} \in \Omega(M')$.

5) The *Lie derivative* L_X doesn't change the degree of a differential form. For $\alpha^{(q)} \in \Omega^q(M)$ the Lie derivative $L_X\alpha^{(q)}$ is defined at every point $m \in M$ by

$$(L_X\alpha^{(q)})_m := \frac{d}{dt}\left((F_X(t)^* \alpha^{(q)})_m\right|_{t=0}.$$

Here F_X means the *flow* associated to $X \in V(M)$. Grossly stated, this is a one–parameter family $F_X(t)$ of diffeomorphisms *in the direction* X.

Somewhat more precisely, a fundamental statement, which will be heavily used in what comes next, and which can be proven with the help of the theorem for the solution of systems of ordinary differential equations (see ABRAHAM–MARSDEN ([**AM**], p. 67) or STERNBERG ([**St**], p. 90)), says that the integral curves γ_X to a differentiable vector field X with $\dot{\gamma}_X(0) = X_{m'}$ for all $m' = \gamma_X(0)$ from a neighborhood U of m form a flow $F = F_X$; that is, a family $(F_X(t))_t$ of diffeomorphisms from U to $F(U)$, where the parameter $t \in \mathbb{R}$ is taken from an interval containing 0. Next for the *complete* X for which $U = M$ can be taken, later also made more general (see ABRAHAM–MARSDEN ([**AM**], p. 90)) and thus made independent of U, we can make the following definition:

$$L_X \alpha^{(q)} = \frac{d}{dt} (F_X(t)^* \alpha^{(q)})\Big|_{t=0}.$$

For these L_X we have the computation rules (see ABRAHAM–MARSDEN ([**AM**], pp. 113 ff.))

i) $L_X d\alpha = dL_X\alpha \quad \text{for } \alpha \in \Omega(M)$,

ii) $L_X\alpha = d(i(X)\alpha) + i(X)d\alpha$,

iii) $L_X\alpha$ is $\mathbb{R}$–bilinear in X and α with $L_X(\alpha\wedge\beta) = L_X\alpha\wedge\beta+\alpha\wedge L_X\beta$,

iv) $L_{fX}\alpha = fL_X\alpha + df \wedge i(X)\alpha \quad \text{for } f \in \mathcal{F}(M)$,

v) $L_X i(X)\alpha = i(X)L_X\alpha$.

In complete analogy, one can define the Lie derivative for a vector field $Y \in V(M)$ as

$$L_X Y := \frac{d}{dt} (F_X(-t)_* Y)\Big|_{t=0}.$$

This is, like the definition of $L_X\alpha^{(q)}$ above, a kind of symbolic notation expressing the derivative in the direction of the vector field f_X. In principle one has to go back to the definition of maps of cotangent and tangent spaces from the end of Section A.2, and arrive at

$$(L_X Y)_m = \frac{d}{dt} \left((F_X(t)_*^{-1})_n Y_n\right)\big|_{t=0} \quad \text{for} \quad n := F_X(t)m$$

and

$$(L_X \alpha)_m = \frac{d}{dt} (F_X(t)_m^* \alpha_n)|_{t=0} .$$

This will be used further in Section B.1, and more details can be found in ABRAHAM–MARSDEN ([**AM**], p. 90/1).

There is yet another important concept of derivative, the *covariant derivatives*. However, these are only definable in the context of *connections*, which will now be introduced in a general context.

A.5. Connections

In [**Ch**], p. 33, Chern introduces the general concept of a connection on a differentiable vector bundle E over a manifold M. Here we only consider the case of the tangent bundle TM on M (see Arnold [**A**] and Abraham–Marsden [**AM**]). In what follows we give a formulation of under what conditions a differentiable vector field $X \in V(M)$, viewed as a differentiable global section of TM over M, can be understood as consisting of *parallel* vectors. Or – possibly easier to visualize – how a vector $X_m \in T_mM$ along a curve $\gamma : J \to M$ passing through m can be *displaced parallelly*. This can be accomplished by making this image more precise. That is, between the tangent spaces T_mM and $T_{m'}M$ of two neighboring points m and m', which are isomorphic only through an abstract isomorphism, we look for a *connection*, i.e. a concrete isomorphism $\phi_{m,m'}$ depending differentiably on (m, m') and such that $\phi_{m,m} = id$. One way to make this more precise consists of defining a map for a given vector field $X \in V(M)$ at every point $m \in M$, which to every $X_m \in T_mM$ and every *direction* $Y_m \in T_mM$ assigns a derivative $(\nabla_Y X)_m$ of X_m in the direction Y_m. X_m does not change in the direction fixed by Y_m if this derivative $(\nabla_Y X)_m$ is 0. This then leads to the following definition (see Abraham–Marsden ([**AM**], p. 145)).

Definition A.35. A *connection* ∇ on a differentiable manifold M is a map

$$\begin{aligned} \nabla : V(M) \times V(M) &\longrightarrow V(M), \\ (X, Y) &\longmapsto \nabla_Y X, \end{aligned}$$

satisfying

a) ∇ is $\mathbb{R}$*–bilinear in* $X, Y \in V(M)$,

b) $\nabla_{fY} X = f\nabla_Y X$, and $\nabla_Y(fX) = f\nabla_Y X + (L_Y f)X$ for all $f \in \mathcal{F}(M)$.

$\nabla_Y X$ is called a *covariant derivative* of X along Y (to the connection ∇).

$L_Y f$ means the Lie derivative of f, as covered in the last section. In a chart (φ, U), with coordinates x and the usual basis $e_i = \frac{\partial}{\partial x_i}$ for T_mM with $m \in U$, let the *Christoffel symbol* Γ^i_{jk} *of the connection* ∇ be introduced as

$$\nabla_{e_j} e_k = \sum_{i=1}^{p} \Gamma^i_{jk} e_i.$$

For $X = \sum a_i e_i$ and $Y = \sum b_j e_j$ we then have

(1) $$(\nabla_Y X) = \sum_i c_i e_i \text{ with } c_i = \sum_j \frac{\partial a_i}{\partial x_j} b_j + \sum_{j,k} \Gamma^i_{jk} a_k b_j.$$

Now let $\gamma : J \to M$ be a curve in M. Unfortunately the associated tangent vectors $\dot{\gamma}(t) \in T_m M$ for $m = \gamma(t)$ do not form a vector field Y on M, but it still makes sense to see these as a part of such a vector field and to denote by $\nabla_{\dot{\gamma}(t)} X$ the *covariant derivative of* X *along* γ. We get

DEFINITION A.36. The vector field $X \in V(M)$ consists of parallel vectors *along* γ when

$$\nabla_{\dot{\gamma}(t)} X = 0 \text{ for all } t \in J.$$

With the use of the frequently needed notation

$$X = \sum X_i e_i$$

we have, from (1), for the covariant derivative of X along γ

(2) $$(\nabla_{\dot{\gamma}(t)} X)_i = \frac{d}{dt}(X_i(\gamma(t))) + \sum_{j,k} \Gamma^i_{jk}(\gamma(t)) X_k(\gamma(t)) \dot{\gamma}_j(t) \; (i = 1, \dots, p).$$

When now the tangent vectors $\dot{\gamma}(t)$ to the curve γ are substituted for X, we get, from (2), as a condition for the tangent vectors being parallel in the sense of the direction fixed by the connection ∇, the system of differential equations

(3) $$\ddot{\gamma}_i(t) + \sum_{j,k} \Gamma^i_{j,k}(\gamma(t)) \dot{\gamma}_j(t) \dot{\gamma}_k(t) = 0.$$

The most important special case is when the connection is fixed and M has the structure of a Riemannian (or pseudo–Riemannian) manifold and then one is given a metric fundamental tensor $g = (g_{ik})$, respectively, a symmetric 2–form

$$ds^2 = \sum g_{ik} dx_i dx_k$$

Then the solution curve γ of the differential equation (3) is called a *geodesic*. It is a fundamental fact that for such a curve the *length of* γ,

$$l(\gamma) = \int_\gamma \sqrt{ds^2},$$

has an extreme value. For the given g_{ik}, respectively,

$$(g^{ik}) := {}^t g^{-1} = (g_{ki})^{-1},$$

we calculate the Christoffel symbol of the connections to g as

$$\Gamma^k_{ij} = (1/2) \sum_l (\partial_i g_{jl} + \partial_j g_{il} - \partial_l g_{ij}) g^{lk}.$$

Connections on vector bundles. Inspection of formula (1) shows that in the definition of connections given above, the coordinates b_j in the *direction* Y' remain unchanged under the covariant derivative. As in CHERN ([**Ch**], p. 33), one can give a definition for a connection in the general situation of a differentiable vector bundle E of rank d over M with $\Gamma(E)$ as the space of global sections over M.

DEFINITION A.37. A *connection* ∇ on $E \xrightarrow{\pi} M$ is a map

$$\nabla : \Gamma(E) \to \Gamma(T^*M \otimes E)$$

with

i) $\nabla(\xi + \eta) = \nabla\xi + \nabla\eta$ for all $\xi, \eta \in \Gamma(E)$,

ii) $\nabla(f\xi) = df \otimes \xi + f\nabla\xi$ for all $\xi \in \Gamma(E), f \in \mathcal{F}(M)$.

Locally on an open set $U \subset M$ with a *d–frame* $s_1, \ldots, s_d$ (that is, sections $s_i \in \Gamma(E, U)$ so that $s_1(m), \ldots, s_d(m)$ are linearly independent for all $m \in U$) this is written as

$$\nabla s_i = \sum_{j=1}^{d} \omega_i^j s_i, \quad i = 1, \ldots, d,$$

or in matrix notation as $\nabla s = \omega s$, where the ω_i^j are 1–forms on U. For

$$\xi \in \Gamma(E, U), \text{ thus } \xi = \sum \xi_i s_i \text{ with } \xi_i \in \mathcal{F}(U),$$

we then have because of i) and ii)

$$\nabla\xi = \sum_i (d\xi_i + \sum_j \xi_i \omega_j^i) s_i. \tag{4}$$

Thus

$$d\xi_i + \sum_{j=1}^{d} \xi_j \omega_j^i = \sum_{k=1}^{p} (\frac{\partial \xi_i}{\partial x_k} + \sum_{j=1}^{d} \Gamma_{jk}^i \xi_j) dx_k, \qquad i = 1, \ldots, d. \tag{5}$$

For the case $E = TM$, the equations (4) and (5) can be equated with equation (1), and then it can be seen that Definition A.36 is carried into Definition A.37 when for $X, Y \in V(M) = \Gamma(TM)$ one substitutes

$$\nabla_Y X = (\nabla X, Y),$$

where here $(\ \ ,\ \ ) : \Gamma(T^*M \otimes TM) \times V(M) \to V(M)$ is induced by the standard pairing

$$dx_i \left(\frac{\partial}{\partial x_j} \right) = \delta_{ij}.$$

The generalization of Definition A.36 of connections makes it possible to make sense of a vector field X being parallel to a curve. Thus, as a consequence of Definition A.37, a section $\xi \in \Gamma(E)$ will be called *horizontal* when it satisfies

$$\nabla \xi = 0,$$

or, locally,

$$d\xi_i + \sum_{j=1}^{d} \omega_j^i \xi_j = 0, \qquad i = 1, \ldots, d.$$

This can be understood as a system of partial differential equations. In general, they will have no solution, but carry over to a solvable system of ordinary differential equations, when one searches for a horizontal section over a given curve γ on M.

Up to this point we have considered only one chart φ and a fixed d–frame s_i. Suppose now that another d–frame is given by

$$s' = As,$$

where A is a nonsingular $d \times d$ matrix of differentiable functions on U. If ω' relative to s' is given by

$$\nabla s' = \omega' s',$$

we derive, as the transformation law of connections,

$$\omega' A = dA + A\omega \tag{6}$$

We further introduce a $d \times d$ matrix Ω of 2–forms, called the *curvature matrix* relative to s, according to the formula

$$\Omega := d\omega + \omega \wedge \omega.$$

Then we can deduce, after exterior derivation of (6) with the matrix Ω' corresponding to s', that the transformation law is

$$\Omega' A = A\Omega.$$

Here it can be seen that the vanishing of Ω is independent of the choice of the d–frame. A connection ω with

$$\Omega = 0$$

is called *flat* or *integrable.* When the connection is derived from a metric, the curvature has the geometric meaning that is studied in differential geometry. In any case, one can by the exterior derivation of the defining equation of Ω arrive at the *Bianchi identity*

$$d\Omega + \Omega \wedge \omega - \omega \wedge \Omega = 0.$$

Given a connection ∇ on the tangent bundle TM, as in (4) and (5), with the help of the Christoffel symbols Γ^i_{jk} and the 1–forms written as

$$\omega_i^k = \sum_{j=1}^{p} \Gamma_{ij}^k dx_j,$$

one may now give, in a natural way, definitions for *covariant derivatives* of differential forms, tensors and, more generally, tensors constructed from differential forms (see for example, ABRAHAM–MARSDEN [**AM**], p. 148) and KÄHLER [**K2**], p. 440)). For a differential form $\alpha \in \Omega^{(q)}(M)$ and $h = 1, \ldots, p$, the *covariant derivative in the direction* h is

$$d_h\alpha := \frac{\partial \alpha}{\partial x_h} - \sum_r \omega_h^r \wedge e_r\alpha$$

with

$$e_r\alpha = \beta, \quad \text{when } \alpha = dx_r \wedge \beta + \gamma,$$

where dx_r does not arise in γ; thus $e_r\alpha = i(X_r)\alpha$ with inner multiplication discussed in Section A.4. It is but a short step to generate a $(q+1)$–form by the sum

$$\sum_{h=1}^{p} dx_h \wedge d_h\alpha,$$

which then, on the basis of the symmetry of Γ^i_{jk}, agrees with the exterior differential $d\alpha$, also discussed in Section A.4.

The covariant derivative of a p–fold contra and q–fold covariant tensor t is a $(p, q+1)$–tensor ∇t given by

$$\begin{aligned}(\nabla t)^{i_1\ldots i_p}_{j_1\ldots j_q h} = \frac{\partial}{\partial x_h} t^{i_1\ldots i_p}_{j_1\ldots j_q} &+ \sum_l t^{l i_2\ldots i_p}_{j_1\ldots j_q}\Gamma^{i_1}_{hl} + \ldots + \sum_l t^{i_1\ldots l}_{j_1\ldots j_q}\Gamma^{i_p}_{hl} \\ &- \sum_l t^{i_1\ldots i_p}_{l j_2\ldots j_q}\Gamma^l_{h j_1} - \ldots - \sum_l t^{i_1\ldots i_p}_{j_1\ldots l}\Gamma^l_{h j_p}.\end{aligned}$$

For a much more general treatment which avoids all these indices, the interested reader is referred to *Equations Différentielles à Points Singuliers Réguliers* by DELIGNE [**D**].

Appendix B

Lie Groups and Lie Algebras

As in Appendix A, we shall explicate from this large and important area of mathematics only the major results and concepts needed in symplectic geometry. As a guideline, Chapter 6 of KIRILLOV'S book *Elements of the Theory of Representations* [**Ki**] and also the corresponding sections from ABRAHAM–MARSDEN [**AM**] will serve well. The interested reader might also want to keep at hand a book with Lie groups or Lie algebras in the title. Particularly recommended are the classics of WIGNER [**Wi**], WEYL [**W**] and CHEVALLEY [**Ce**].

B.1. Lie algebras and vector fields

Lie algebras made their first appearance as *infinitesimal Lie groups*, but quickly acquired meaning as self–standing algebraic objects. In the following, the ground field K will be assumed to be $\mathbb{R}$ or $\mathbb{C}$. The theory of Lie algebras can, however, be considered for a completely general field.

DEFINITION B.1. A K vector space $\mathfrak{g}$ is called a *Lie algebra* if a *Lie product* (or a *Lie bracket*) is defined on it; that is, a K–bilinear map

$$[\ ,\ ] : \mathfrak{g} \times \mathfrak{g} \to \mathfrak{g}$$

satisfying

- (Antisymmetry) $[X, X] = 0$ for all $X \in \mathfrak{g}$,
- (Jacobi identity) $[[X, Y], Z] + [[Y, Z], X] + [[Z, X], Y] = 0$ for all $X, Y, Z \in \mathfrak{g}$.

Remark B.2. If $\mathfrak{g}$ is finite–dimensional with basis $(X_1, \ldots, X_n)$, it is clearly enough to know the values of the Lie brackets $[X_i, X_j]$ for pairs of basis elements. In other words, one must know the coefficients c_{ij}^k $(i, j, k = 1, \ldots, n)$ from the expressions

$$[X_i, X_j] = \sum_{k=1}^{n} c_{ij}^k X_k.$$

These coefficients are called the *structure constants* of $\mathfrak{g}$ (relative to the given basis $(X_1, \ldots, X_n)$).

EXAMPLE B.3.

(1) The Lie algebra $\mathfrak{g}$ is called *abelian* or *commutative* whenever

$$[X, Y] = 0 \quad \text{for all} \quad X, Y \in \mathfrak{g}.$$

Every K vector space can be considered as a commutative Lie algebra when equipped with this trivial Lie bracket.

(2) Every associative K–algebra $\mathbf{A}$ can be taken to be a Lie algebra, by taking for the Lie bracket

$$[X, Y] := XY - YX \quad \text{for} \quad X, Y \in \mathbf{A}.$$

In this manner, for the space of all $p \times p$ matrices $\mathbf{A} = M_p(K)$ we have the Lie algebra called $\mathfrak{gl}_p(K)$.

(3) Every (not necessarily associative) K–algebra $\mathbf{A}$ allows the construction of a Lie algebra Der $\mathbf{A}$, given by the K–linear maps called derivations; that is, maps satisfying

$$D : A \to A, \text{ with } D(XY) = (DX)Y + XDY \quad \text{for all } X, Y \in \mathbf{A}$$

and $[D_1, D_2] := D_1 D_2 - D_2 D_1$ as Lie bracket. Der $\mathbf{A}$ is called the algebra of *derivations*, or also *differentiations*, on $\mathbf{A}$.

By an easy refinement of Theorem A.13, it can be shown (see ABRAHAM–MARSDEN ([**AM**], p. 83)) that:

THEOREM B.4. *The space $V(M)$ of differentiable vector fields on a differentiable manifold M is isomorphic to* Der $(\mathcal{F}(M))$ *as K vector spaces.*

This isomorphism is given by assigning to a vector field X on M the derivation L_X defined by

$$L_X f(m) = \langle X_m, (df)_m \rangle \text{ for } f \in \mathcal{F}(M) \text{ and } m \in M.$$

$L_X f$ will be called the *Lie derivative of f relative to X*, and has already been introduced at the end of Section A.3 as well as in the discussion of the definition of tangent spaces in Section A.1, where it was written, because of local considerations, as $L_X f(m) = L_{X_m} f$.

The isomorphism of the theorem gives a Lie algebra structure on $V(M)$, where $[X, Y]$ for $X, Y \in V(M)$ is the uniquely defined vector field given by

$$L_{[X,Y]} = [L_X, L_Y].$$

For a diffeomorphism $F : M \to M'$; in accordance with the conclusions from Section A.3, $F_* X \in V(M')$ is taken as the image of $X \in V(M)$. The map F_* is compatible with the Lie bracket (see ABRAHAM–MARSDEN ([**AM**], p. 85)).

Remark B.5. F_* is a *Lie algebra homomorphism*; that is, F_* is K–linear with

$$F_*[X, Y] = [F_* X, F_* Y] \quad \text{for} \quad X, Y \in V(M).$$

The Lie derivative of a vector field. It is not very easy to prove the statement that the map defined in terms of flows, F_X to $X \in V(M)$, at the conclusion of Section A.4,

$$\begin{aligned} L_X : V(M) &\longrightarrow V(M), \\ Y &\longmapsto L_X Y := \frac{d}{dt} (F_X(-t)_* Y)\Big|_{t=0}, \end{aligned}$$

is the same map given in terms of the Lie bracket as

$$L_X Y = [X, Y].$$

B.2. Lie groups and invariant vector fields

DEFINITION B.6. A *Lie group* G is a finite–dimensional manifold, on which a group structure is defined, whereby the group operations are given by differentiable maps; that is, the maps

$$\begin{aligned} G \times G &\longrightarrow G, & \text{and} \quad G &\longrightarrow G, \\ (g, h) &\longmapsto gh, & g &\longmapsto g^{-1}, \end{aligned}$$

are differentiable.

Here, we will focus on the case of real Lie groups; analogous constructs can be formed over the complex numbers.

EXAMPLE B.7.

(1) $G = \mathbb{R}^n$ with addition as product.

(2) $G = \mathbb{R}_{>0}$ with multiplication as product; or analogously,

$$G = S^1 = \{z \in \mathbb{C}, |z| = 1\}.$$

(3) $G = GL(n, \mathbb{R})$ with matrix multiplication as product.

(4) $G = G_1 \times G_2$, where both G_1 and G_2 are Lie groups.

(5) $G = H(\mathbb{R}) = \{h = (\lambda, \mu, \kappa) \in \mathbb{R}^3\}$, the *Heisenberg group of degree* 1 with the product

$$hh' = (\lambda + \lambda', \mu + \mu', \kappa + \kappa' + \lambda\mu' - \lambda'\mu).$$

Every Lie group G gives rise to a Lie algebra $\mathfrak{g} = \text{Lie } G$, which can be thought of as the *linearization* of G and whose structure is fixed by G, here in a neighborhood of the unity. For this assignment there are several equivalent constructions. Here we take as the starting point a concept which also finds many other uses.

NOTATION B.8. For each $g_0 \in G$ denote by λ_{g_0} the function defined by

$$g \mapsto \lambda_{g_0} g := g_0 g,$$

the so-called *left translation* given by g_0, and by ρ_{g_0} the *right translation* given by

$$g \mapsto \rho_{g_0} g := g g_0.$$

For every g_0, the maps λ_{g_0} and ρ_{g_0} are diffeomorphisms of G to itself. They can also be used to *shift* given vector fields X on G. This makes possible the concepts of right–invariant and left–invariant vector fields on G.

DEFINITION B.9. $X \in V(G)$ is called *left–invariant* whenever

$$(\lambda_{g_0})_* X = X \quad \text{for all} \quad g_0 \in G.$$

The space $V_l(G)$ of left–invariant vector fields on G is clearly a vector subspace of $V(G)$, which can then be proved to be a Lie subalgebra of the Lie algebra $V(G)$ (and therefore is carried over to $\mathfrak{g} = \operatorname{Lie} G$).

Remark B.10. The tangent space T_eG to G at the unity element $e \in G$ is isomorphic to $V_l(G)$ as a vector space:

$$V_l(G) \underset{\varphi_2}{\overset{\varphi_1}{\rightleftarrows}} T_eG.$$

Here φ_1 is given by

$$\varphi_1(X) := X_e \quad \text{for} \quad X \in V_l(G)$$

and φ_2 by

$$\varphi_2(\xi) := X \quad \text{with} \quad X_g := (\lambda_g)_* \xi \quad \text{for} \quad \xi \in T_eG.$$

Then we clearly have

$$\varphi_1 \varphi_2(\xi) = \varphi_1((\lambda_g)_* \xi) = (\lambda_e)_* \xi = \xi.$$

Moreover, from the left–invariance of X

$$(\varphi_2\varphi_1(X))_g = (\lambda_g)_* X_e = X_g;$$

thus $\varphi_1\varphi_2 = id_{T_eG}$ and $\varphi_2\varphi_1 = id_{V_l(G)}$.

Remark B.11. If $X, Y \in V_l(G)$, then $[X, Y]$ is also in $V_l(G)$. Then, because of Remark B.2, we have

$$(\lambda_g)_*[X, Y] = [(\lambda_{g*})X, (\lambda_g)_*Y] = [X, Y].$$

And this allows us to fix a definition:

DEFINITION B.12. The vector space T_eG with the Lie algebra structure induced by the isomorphism with $V_l(G)$ is called the *Lie algebra* $\mathfrak{g} = \mathrm{Lie}\, G$ of G.

EXAMPLE B.13.

$$\begin{aligned} \mathfrak{gl}_n(\mathbb{R}) := \mathrm{Lie}\, GL_n(\mathbb{R}) &= M_n(\mathbb{R}), \\ \mathfrak{sl}_n(\mathbb{R}) := \mathrm{Lie}\, SL_n(\mathbb{R}) &= \{X \in M_n(\mathbb{R}),\ \ \mathrm{trace}\, X - 0\}, \\ \mathfrak{sp}_n(\mathbb{R}) := \mathrm{Lie}\, Sp_n(\mathbb{R}) &= \{X \in M_{2n}(\mathbb{R}),\ \ {}^tXJ + JX = 0\}, \end{aligned}$$

as well as

$$\mathfrak{o}\,(3) := \mathrm{Lie}\, O(3) \simeq \mathbb{R}^3$$

with

$$\begin{aligned} X &\longleftrightarrow \xi, \\ [X, Y] &\longleftrightarrow \xi \times \eta \quad \text{(the vector product in } \mathbb{R}^3). \end{aligned}$$

B.3. One–parameter subgroups and the exponent map

As already indicated, the passage from a Lie group G to its Lie algebra $\mathfrak{g}$ can, to a certain degree, be reversed. To make this statement more precise requires the following concepts, which can only be briefly mentioned here.

To each $\xi \in T_eG$ denote by X its associated left–invariant vector field (with $X_e = \xi$). Then denote by

$$\begin{aligned} \gamma_\xi : \mathbb{R} &\longrightarrow G, \\ t &\longmapsto \exp t\xi, \end{aligned}$$

the *integral curve* to X which for $t = 0$ goes through e and whose tangent vector $\dot\gamma_\xi(t)$ at every point $\gamma_\xi(t)$ is equal to $X_{\gamma_\xi(t)}$. That such a curve exists is, as already discussed at the end of Section B.1, a consequence of the existence and uniqueness theorem for solutions of systems of ordinary differential equations (and a simple corollary). For such a curve, it can be shown (see, for example, ABRAHAM–MARSDEN ([**AM**], p. 255)) that

$$\exp{(t+s)\xi} = \exp{t\xi} \cdot \exp{s\xi};$$

that is,

$$\gamma_\xi : \mathbb{R} \to G$$

is a (differentiable) group homomorphism. γ_ξ is called a *1–parameter subgroup* of G. This now makes possible the next definition.

DEFINITION B.14. The map

$$\begin{array}{rcl} \exp : T_eG & \longrightarrow & G, \\ \xi & \longmapsto & \gamma_\xi(1) = \exp \xi, \end{array}$$

is called the *exponential map* of the Lie algebra $\mathfrak{g} = T_eG$ in G.

This map turns out to be differentiable and induces the identity map on the tangent space $T_0(T_eG) \simeq T_eG$. Therefore it is a local diffeomorphism, but not a diffeomorphism on G.

Remark B.15. For a differentiable homomorphism $F : H \to G$ between two Lie groups H and G and for the induced map

$$T_eF := (F_*)_e : T_eH \to T_eG$$

we have the commutation rule

$$F(\exp_H \eta) = \exp_G (T_eF)\eta \quad \text{for all} \quad \eta \in T_eH = \operatorname{Lie} H.$$

Since the map

$$\gamma : \mathbb{R} \to F(\exp_H t\eta)$$

is a 1–parameter subgroup of G, it is of the form

$$\gamma(t) = \exp_G t\xi \quad \text{with} \quad \xi = \frac{d}{dt}\gamma(t)|_{t=0} = (T_eF)(\eta);$$

that is,

$$F(\exp_H \eta) = \gamma(1) = \gamma_\xi(1) = \exp_G \xi = \exp_G (T_eF)\eta.$$

An important *special case* is given by conjugation:

$$\begin{array}{rcl} \kappa_g : G & \longrightarrow & G, \\ h & \longmapsto & ghg^{-1} = \rho_{g^{-1}}\lambda_g h \quad \text{for} \quad g \in G. \end{array}$$

This is a differentiable map, and actually an *inner automorphism* of G. Now denote by

$$Ad_g := T_e\kappa_g = T_e(\rho_g^{-1}\lambda_g) : T_eG \to T_eG$$

the *adjoint map* associated to $g \in G$. As a consequence of the last remark, it follows that

$$\exp(Ad_g\,\xi) = \kappa_g \exp\xi = g\,(\exp\xi)\,g^{-1} \quad \text{for all } \xi \in T_eG \text{ and } g \in G.$$

EXAMPLE B.16.

(1) $G = \mathbb{R}^n$, with addition as product, has $\mathfrak{g} = \operatorname{Lie} G = \mathbb{R}^n$, and exp: $\mathbb{R}^n \to \mathbb{R}^n$ is the identity.

(2) For $G = GL_n(\mathbb{R})$ and its subgroups, the exponential map is the generalization of the exponential function of matrices; thus

$$\begin{aligned} \exp : M_n(\mathbb{R}) &\longrightarrow GL_n(\mathbb{R}), \\ A &\longmapsto \sum_{m=0}^{\infty} A^m/m!\,. \end{aligned}$$

To every $A \in M_n(\mathbb{R})$ there belongs the 1–parameter subgroup γ_A, given by

$$\gamma_A(t) = \exp tA = \sum_{m=0}^{\infty} t^m A^m/m!$$

From this, for $C \in GL_n(\mathbb{R})$, we derive the frequently used computation rule

$$\exp\,(CAC^{-1}) = C(\exp A)C^{-1}.$$

Appendix C

A Little Cohomology Theory

Homology and cohomology groups are important tools for both describing and characterizing objects in almost every area of mathematics. They are introduced under the topics of *homological algebra* and/or *algebraic topology.* Here we will only give a few definitions from various sources (for instance, KIRILLOV [**Ki**], and GUILLEMIN–STERNBERG [**GS**]) and treat enough of their elementary properties to suffice for our study of symplectic geometry. For a systematic introduction, we recommend MACLANE'S book *Homology* [**ML**] or GODEMENT'S book *Topologie algébrique et théorie des faisceaux* [**Go**], which has retained its value as a classical introduction to sheaves and their cohomology.

C.1. Cohomology of groups

Let G and M be groups with M abelian, and let G *operate on M from the left*, that is, there is a map

$$\begin{aligned} G \times M &\longrightarrow M, \\ (g, m) &\longmapsto gm, \end{aligned}$$

with

$$\begin{aligned} (gg')m &= g(g'm), \\ em &= m, \\ g(m + m') &= gm + gm' \end{aligned}$$

for all $g, g' \in G$ and $m, m' \in M$. Then (see KIRILLOV ([**Ki**], p. 21)) an *n–dimensional cochain* c is an $(n+1)$–linear map

$$c : \underbrace{G \times \ldots \times G}_{n+1} \longrightarrow M$$

with

$$c(gg_0, \ldots, gg_n) = gc(g_0, \ldots, g_n) \quad \text{for all} \quad g, g_0, \ldots, g_n \in G.$$

The collection of all n–dimensional (or, more simply, n-) cochains is a group, and will be denoted by

$$C^n(G, M).$$

Then a *coboundary operator d*

$$\begin{aligned} d : C^n(G, M) &\longrightarrow C^{n+1}(G, M), \\ c &\longmapsto dc, \end{aligned}$$

is given by

$$dc(g_0, \ldots, g_{n+1}) = \sum_{i=0}^{n+1} (-1)^i c(g_0, \ldots, \hat{g}_i, \ldots, g_{n+1}).$$

If

$$c = db,$$

then $c \in C^n(G, M)$ is called a *coboundary of the cochain* $b \in C^{n-1}(G, M)$; and c is called a *cocycle*, if

$$dc = 0.$$

Then we get in C^n the subgroup of cocycles Z^n and the subgroup of coboundaries B^n. The following statement is central.

Remark C.1. We have

$$d \circ d = 0.$$

EXERCISE C.2. Prove the remark.

We thus have

$$C^n(G, M) \supset Z^n(G, M) \supset B^n(G, M),$$

and we can form the factor group

$$H^n(G, M) := Z^n(G, M)/B^n(G, M).$$

This is again a group, and is called the n–th *cohomology group of the group G with coefficients in M*. Taking a direct sum, we obtain

$$H^*(G, M) := \bigoplus_n H^n(G, M).$$

This is a graded ring or a graded algebra, if M is not only a group, but a ring or an algebra as well.

For practical computation, the cochain functions c can also be replaced by $\tilde{c}$:

$$\tilde{c}(h_1, \ldots, h_n) := c(e, h_1, h_1 h_2, \ldots, h_1 \ldots h_n).$$

For these $\tilde{c}$ the coboundary operator is then

$$d\tilde{c}(h_1, \ldots, h_{n+1}) = h_1 \tilde{c}(h_2, \ldots, h_{n+1}) + \sum_{i=1}^{n} (-1)^i \tilde{c}(h_1, \ldots, h_{i-1}, h_i h_{i+1}, h_{i+2}, \ldots, h_{n+1}).$$

EXERCISE C.3. Prove this last formula.

As a hint as to the usefulness of this concept, we inform the reader of the following example. Given two groups G_0 and G_1 with an operation of G_1 on $Z(G_0) = \{g \in G_0, gg_0 = g_0 g, g_0 \in G_0\}$, the *center* of G_0, one can identify $H^2(G_1, Z(G_0))$ with the set of equivalence classes of central extensions

$$1 \longrightarrow G_0 \longrightarrow G \longrightarrow G_1 \longrightarrow 1$$

of G_1 by G_0 (see KIRILLOV ([**Ki**], p. 18)).

C.2. Cohomology of Lie algebras

Let $\mathfrak{g}$ be a Lie algebra and V a $\mathfrak{g}$–module (see GUILLEMIN–STERNBERG ([**GS**], p. 417)). Then denote by $C^k(\mathfrak{g}, V)$ the collection of all n–cochains, that is, the antisymmetric n–linear maps

$$f : \mathfrak{g} \times \ldots \times \mathfrak{g} \to V.$$

Then take as coboundary operator the operator

$$\delta : C^k(\mathfrak{g}, V) \to C^{k+1}(\mathfrak{g}, V)$$

defined by

$$\begin{aligned} \delta f(\xi_0, \ldots, \xi_k) &:= \sum_{i=0}^{k} (-1)^i \xi_i f(\xi_0, \ldots, \hat{\xi}_i, \ldots, \xi_k) \\ &+ \sum_{i<j} (-1)^{i+j} f([\xi_i, \xi_j], \xi_0, \ldots, \hat{\xi}_i, \ldots, \hat{\xi}_j, \ldots, \xi_k). \end{aligned}$$

Thus, with this arrangement, we have for $k = 0$

$$\delta f(\xi) = \xi f,$$

for $k = 1$

$$\delta f(\xi_0, \xi_1) = \xi_0 f(\xi_1) - \xi_1 f(\xi_0) - f([\xi_0, \xi_1]),$$

and for $k = 2$

$$\begin{aligned}\delta f(\xi_0, \xi_1, \xi_2) &= \xi_0 f(\xi_1, \xi_2) - \xi_1 f(\xi_0, \xi_2) + \xi_2 f(\xi_0, \xi_1)\\ &\quad - f([\xi_0, \xi_1], \xi_2) + f([\xi_0, \xi_2], \xi_1) - f([\xi_1, \xi_2], \xi_0).\end{aligned}$$

EXERCISE C.4. Verify that $\delta^2 = 0$.

Thus, in analogy to Section C.1, we may define the cohomology groups

$$H^k(\mathfrak{g}, V) := Z^k(\mathfrak{g}, V)/B^k(\mathfrak{g}, V).$$

In the special case that V is a trivial $\mathfrak{g}$–module (that is, $\xi v = 0$ for all $\xi \in \mathfrak{g}$ and $v \in V$), the first terms of the coboundary operators do not appear, and if V is simply K, we write

$$H^k(\mathfrak{g}) := H^k(\mathfrak{g},\, K).$$

The vanishing of $H^1(\mathfrak{g})$ and $H^2(\mathfrak{g})$ allows for important reductions in symplectic geometry (see, for example, the theorem of Kostant and Souriau in GUILLEMIN–STERNBERG ([**GS**], p. 179), which is covered in Section 2.5). Here let us but list the following general statements from GUILLEMIN–STERNBERG ([**GS**], pp. 418 ff.)):

(1) $H^1(\mathfrak{g}, V) = \{0\}$ for all V precisely when every representation of $\mathfrak{g}$ is completely reducible.

(2) If the above is satisfied, it follows that $H^2(\mathfrak{g}) = \{0\}$.

(3) If $\mathfrak{g}$ is semisimple (that is, it has no commutative ideals), we have, as in 1), that $H^1(\mathfrak{g}, V) = 0$ for all V.

C.3. Cohomology of manifolds

It is especially important to actually compute homology and cohomology groups for topological and geometric objects, for example *manifolds, varieties, schemes*, etc. This has led to a whole art of technical apparati making such computations possible. Here we give a brief treatment of the following related concepts (see KIRILLOV ([**Ki**], p. 9)).

Let G be an abelian group, M a manifold, and $\mathfrak{U} = \{U_\alpha\}_{\alpha \in I}$ a covering of M by open sets. A *k–cochain of $\mathfrak{U}$ with coefficients in G* is then a skew–symmetric map c defined on every $(k+1)$–tuple $(\alpha_0, \dots, \alpha_k) \in I^{k+1}$ with

$$U_{\alpha_o} \cap \ldots \cap U_{\alpha_k} \neq \emptyset$$

to G. Thus

$$c(\dots, \alpha_i, \dots, \alpha_j, \dots) = -c(\dots, \alpha_j, \dots, \alpha_i, \dots).$$

$C^k(\mathfrak{U}, G)$ denotes the group of these k–cochains. The coboundary map

$$d : C^k(\mathfrak{U}, G) \longrightarrow C^{k+1}(\mathfrak{U}, G)$$

is given by

$$dc(\alpha_0, \ldots, \alpha_{k+1}) = \sum_{i=0}^{k+1} (-1)^i c(\alpha_0, \ldots, \hat{\alpha}_i, \ldots, \alpha_{k+1}).$$

We then have that $d^2 = 0$, and therefore, in complete analogy to what has come before, we define

$$H^k(\mathfrak{U}, G) := Z^k(\mathfrak{U}, G)/B^k(\mathfrak{U}, G)$$

as the *Čech cohomology groups of the open covering* $\mathfrak{U}$. This shows a way to give M itself cohomology groups. Namely given a finer open covering $\mathfrak{U}'$ of $\mathfrak{U}$, we get, in a natural way, homomorphisms

$$\tau_{\mathfrak{U}}^{\mathfrak{U}'} : H^k(\mathfrak{U}, G) \longrightarrow H^k(\mathfrak{U}', G).$$

This construct allows us to define *inductive limits* over a directed system of open coverings:

$$H^k(M, G) := \varinjlim_{\mathfrak{U}} H^k(\mathfrak{U}, G).$$

Fortunately these cohomology groups can already be computed from a single covering of $\mathfrak{U}$ (LERAY's Theorem):

$$H^k(M, G) = H^k(\mathfrak{U}, G),$$

when $\mathfrak{U} = \{U_\alpha\}$ and the sets U_α and their intersections all have trivial cohomology in the dimensions $n \geq 1$.

A completely different viewpoint allows one to attach cohomology groups to a real or complex $\mathcal{C}^\infty$-manifold M with the help of differential forms (see Section A.4). For $K = \mathbb{R}$ or $\mathbb{C}$, denote by $Z^k(M, K)$ the closed K–valued k–forms on M; thus $\omega \in \Omega^k(M)$ with $d\omega = 0$. Denote by $B^k(M, K)$ the boundaries of ω, thus those for which there is a $\vartheta \in \Omega^{k-1}(M)$ such that $d\vartheta = \omega$. Then the *de Rham cohomology groups* $H^k_{DR}(M, K)$ are defined by

$$H^k_{DR}(M, K) = Z^k(M, K)/B^k(M, K).$$

De Rham's theorem says that these cohomology groups are the same as the above–defined Čech cohomology groups:

$$H^k_{DR}(M, K) = H^k(M, K).$$

Appendix D

Representations of Groups

Many of the concepts from representation theory have already played a role in the previous appendices. Here, we collect together in one place all sorts of important concepts which are needed in connection with symplectic geometry. As a guideline to this material we take §7 from KIRILLOV'S book [**Ki**]. An introduction to the parts of the theory most relevant to us is also offered in LANG'S book [**L4**].

D.1. Linear representations

Let G be a group and V a K vector space. Here again we are mostly thinking about $K = \mathbb{R}$ or $\mathbb{C}$, although much of what we will say holds also for a general field. We call T a *representation of G in V* when T is a homomorphism of G in Aut V, i.e. when

$$T(g_1 g_2) = T(g_1)\, T(g_2) \quad \text{for } g_1, g_2 \in G.$$

$\dim V$ is taken as the *dimension of T*. Two representations T and T' of G in V and V', respectively, are called *equivalent* when there is a bijective *intertwining operator* between the two; that is, a linear isomorphism $U : V \to V'$ with

$$UT(g) = T'(g)U \quad \text{for all } g \in G.$$

The set of intertwining operators U of V and V' will also be designated by $\mathcal{C}(T, T')$. A representation T is called *reducible* if the representation space V properly contains a non–trivial subspace V_1 invariant under all $T(g)$ with $g \in G$. The restriction of $T(g)$ to V_1 then defines a representation T_1 of G on V_1, which will be called a *subrepresentation of T*. There is then also a natural representation of G on the quotient space V/V_1, which is then called the *quotient representation* (in KIRILLOV ([**Ki**], p. 110), this is called a *factor representation of* T). If the invariant subspace $V_1 \subset V$ allows for an invariant subspace complement T_2, then the representation T on V is called *decomposable* and is written as $T = T_1 \oplus T_2$.

A representation is called (algebraically) *irreducible* if it contains no non–trivial subrepresentation other than itself. A representation T in V is called *completely reducible* if every invariant subspace of V possesses an invariant complement. This is then equivalent to V being completely decomposable into a sum of irreducible subrepresentations.

One of the major tasks of representation theory is to give a complete list of all isomorphism classes of irreducible representations, or at least of all irreducible unitary representations (see the next section). Another task is to take a given completely reducible representation T and to reduce it; that is, to write it as a direct sum of its irreducible parts T_i along with their *multiplicities*

$$\operatorname{mult}(T_i, T) := \dim \mathcal{C}\,(T_i, T).$$

A *character* χ of G is a homomorphism

$$\chi : G \to \mathbb{C}_1 = \{\zeta \in \mathbb{C},\ |\zeta| = 1\}.$$

For a given representation T of G in V, T^* is called the *contragredient representation* if T^* in V^* is given by

$$g \mapsto T^*(g) := T\,(g^{-1})^*.$$

Under these conditions, it is then a fact that for a map of representations $F : V \to V'$ the canonical map $F^* : (V')^* \to V^*$ with

$$F^*(v'^*)(v) = v'^*\big(F\,(v)\big)$$

again gives a map of representations.

It is also but a simple exercise to show that from representations T_i of G in V_i $(i = 1, 2)$ one can obtain new representations in the following two ways.

i) The *direct sum* $T_1 \oplus T_2$, as a representation of G on $V_1 \oplus V_2$, is given by

$$(T_1 \oplus T_2)(g)(v_1 + v_2) = T_1(g)v_1 + T_2(g)v_2 \quad \text{for } v_1 \in V_1,\ v_2 \in V_2.$$

ii) The *tensor product* $T_1 \otimes T_2$, as a representation of G of the tensor product $V_1 \otimes V_2$, is given by

$$(T_1 \otimes T_2)(g)(v_1 \otimes v_2) = T_1(g)v_1 \otimes T_2(g)v_2 \quad \text{for } v_1 \in V_1, v_2 \in V_2.$$

It is a standard result of representation theory that the tensor product of given irreducible representations can be decomposed into irreducible representations.

Parallel to the concept of linear representations, we discuss the *projective representations.* These are maps $T : G \to \operatorname{Aut} V$ satisfying

$$T(g_1 g_2) = c(g_1, g_2)\, T(g_1) T(g_2) \quad \text{for all } g_1, g_2 \in G,$$

where $c : G \times G \to K$ is a function satisfying the functional equation

$$c(g_1, g_2)\, c(g_1 g_2, g_3) = c(g_1, g_2 g_3)\, c(g_2, g_3) \quad \text{for } g_1, g_2, g_3 \in G,$$

which then insures the equality

$$T(g_1 g_2)\, T(g_3) = T(g_1)\, T(g_2 g_3).$$

Such a projective representation of G on V then induces a representation on the projective space, $\mathbb{P}(V)$ associated to V, which is defined by

$$T(g)(v_\sim) = \big(T(g)\, v\big)_\sim, \quad \text{for } v_\sim \in \mathbb{P}(V).$$

D.2. Continuous and unitary representations

In the applications in this text, we will most often encounter G which are Lie groups, and so we have a differentiable and topological structure. Thus, we will need all types of restrictions on the concept of representations of G. Here we will assume the following data is fixed: T is a *continuous representation*, thus a representation of a topological group G in a topological vector space V (mostly a Hilbert space) with

$$\begin{array}{ccl} G \times V & \longrightarrow & V, \\ (g, v) & \longmapsto & T(g)v \ \text{ is continuous.} \end{array}$$

For such continuous representations, the definitions of the previous section are extended in the appropriate sense. For example, in this sense, T_1 would be a *subrepresentation* of T on $V_1 \subset V$ when V_1 is a closed invariant subspace of V. The extension of the other definitions is entirely analogous.

Unitary representations. A representation π of a (not necessarily topological) group G in the space V is called *unitary* if V is a Hilbert space and the operators $T(g)$ for all $g \in G$ are unitary. Unitary representations are the most important in applications to physics. They are particularly tractable, since every unitary representation is completely reducible. The collection of isomorphism classes of irreducible unitary representations of G is called the *unitary dual* and is denoted by $\widehat{G}$. $\widehat{G}$ can be supplied with a topological structure (see KIRILLOV ([**Ki**], p. 113)).

EXAMPLE D.1. Elementary examples are the following:

i) For an abelian group G, every $\pi \in \widehat{G}$ is one–dimensional.

ii) For a compact group G, every $\pi \in \widehat{G}$ is finite–dimensional.

iii) For $G = \mathbb{R}^n$ we have $\widehat{G} = \mathbb{R}^n$, and for $G = S^1$ we have $\widehat{G} = \mathbb{Z}$.

Schur's lemma says

LEMMA D.2.

a) *If T_1 and T_2 are (algebraically) irreducible representations, then every intertwining operator $A \in \mathcal{C}\,(T_1, T_2)$ is either 0 or invertible.*

b) *A unitary representation π is irreducible precisely when*

$$\dim \mathcal{C}\,(\pi, \pi) = 1.$$

D.3. On the construction of representations

There are several ways to approach the representations of a group, in particular for the unitary dual $\widehat{G}$. Here we give a few hints.

Reduction of the regular representation. For a locally compact topological group, denote by $V = L^2(G)$ the Hilbert space of quadratic integrable functions on V with respect to a suitable measure, $d\mu$ (a *Haar measure*, see, for example, KIRILLOV ([**Ki**], p. 130)). Denote by ϱ the representation $\phi \in L^2(G)$ given by

$$\varrho\,(g_0)\,\phi\,(g) = \phi\,(g g_0) \quad \text{for } g, g_0 \in G$$

the so-called *right regular representation.* Correspondingly, the *left regular representation* λ is given by

$$\lambda\,(g_0)\,\phi\,(g) = \phi\,(g_0^{-1} g).$$

An intertwining operator U between ϱ and λ is given by

$$\phi \overset{U}{\longmapsto} \check{\phi} \quad \text{with} \quad \check{\phi}\,(g) := \phi\,(g^{-1}).$$

In the general case, the question arises as to whether all $\pi \in \widehat{G}$ are subrepresentations of ϱ respectively λ. For compact G, this is the case (this is a consequence of the Peter–Weyl theorem).

A variant. We now treat a variant of this. We work in the following situation. G operates contiuously on a differentiable manifold M; that is, there is given a map

$$\begin{array}{ccc} G \times M & \longrightarrow & M, \\ (g, m) & \longmapsto & gm = \varphi_g(m), \end{array}$$

where all the φ_g, for $g \in G$, are continuous, and both

$$\varphi_{g_1 g_2} = \varphi_{g_1} \varphi_{g_2} \qquad \text{for all } \ g_1, g_2 \in G$$

as well as

$$\varphi_e = \mathrm{id}_M$$

hold. Then $\check{\varphi}$, given by

$$\check{\varphi}\,(g)\,\phi\,(m) = \phi\left(\varphi_{g^{-1}}(m)\right) \quad \text{for } \phi \in \mathcal{F}\,(M),$$

is a representation of G on the space $\mathcal{F}\,(M)$ of all arbitrarily often differentiable functions on M. This theme can now be varied in a myriad of ways. To show the simplest case, let j be an *automorphic factor*; that is, a function

$$\begin{array}{rcl} j : G \times M & \longrightarrow & \mathbb{R}^*, \\ (g, m) & \longmapsto & j\,(g, m), \end{array}$$

satisfying the functional equation

$$j\,(g_1 g_2, m) = j\,(g_1, g_2 m)\, j\,(g_2, m) \quad \text{for all } g_1, g_2 \in G, \ m \in M.$$

Then from $\check{\varphi}_j$ with

$$\check{\varphi}_j(g)\, f\,(m) = \phi\left(g^{-1}(m)\right) j\,(g^{-1}, m)$$

we get another representation of G on $\mathcal{F}\,(M)$.

The standard example in this connection is the operation of $G = SL_2(\mathbb{R})$ on $\mathbb{R}^2$ given by multiplication of the matrix $g \in G$ with the column vector $m \in \mathbb{R}^2$. Then $\check{\varphi}$ with

$$\check{\varphi}\,(g)\, f\,(m) = f\,(g^{-1} m) \quad \text{for } f \in V = \mathbb{R}[q, p]$$

is a representation of G on the polynomial ring $\mathbb{R}[q, p]$ in two variables. Clearly the subspaces of homogeneous polynomials of degree $l \geq 0$ remain invariant and give irreducible representations.

Eigenspace representations. As in the previous example, G operates contiuously on a differentiable manifold M. Let D be a G–invariant differential operator on $\mathcal{F}\,(M)$, and let $\mathcal{E}\,(D, \lambda)$ be the eigenspace of D with eigenvalue λ. Then $\check{\varphi}$, defined as in the previous example, is a representation on $\mathcal{E}\,(D, \lambda)$, called an *eigenspace representation.* These representations have been particularly propagated by HELGASON.

The infinitesimal method (for Lie groups). Let π be a continuous representation of a real Lie group G in a topological vector space V. Then denote by V_ω the space of *analytic vectors* in V, that is, those $v \in V$ for which

$$g \mapsto \pi(g)\,v$$

is a real analytic map. From results of Harish–Chandra and Nelson, V_ω is dense in V. π assigns, as an *infinitesimal representation* $d\pi$, a representation of the Lie algebra $\mathfrak{g}$ of G on V_ω given by

$$d\pi(X)\,v := \frac{d}{dt}\,\pi(\exp tX)v\big|_{t=0} \qquad \text{for } X \in \mathfrak{g},\ v \in V_\omega.$$

(A representation of a Lie algebra in a vector space W is a homomorphism of Lie algebras $\mathfrak{g} \to \operatorname{End} V$, where $\operatorname{End} V$ is a Lie algebra as explained in Section B.1. That $d\pi$ gives such a representation can be shown as an exercise.)

The infinitesimal method consists of these steps, but followed in the reverse order:

a) Determination of the irreducible representations $\widehat{\pi}$ of $\mathfrak{g}$, or of the complexification $\mathfrak{g}_c = \mathfrak{g} \otimes \mathbb{C}$. (This is somewhat easy to do for the Heisenberg algebra $\mathfrak{g} = \mathfrak{h}$ and for $\mathfrak{g} = \mathfrak{sl}_2$. The elements for these were provided in Sections 5.1 and 5.2; the rest is a recommended exercise.)

b) Examination of which of these $\widehat{\pi}$ can be integrated to unitary representations π; that is, the determination of those representations π of G with $d\pi = \widehat{\pi}$.

In important cases it turns out that π is uniquely determined by $d\pi$. If $G = H(\mathbb{R})$ is the Heisenberg group, one may take for π the Schrödinger representation, which was described in Section 5.2. For $G = SL_2(\mathbb{R})$, it is relatively easy to determine the unitary dual $\widehat{G}$ (see, for example, Lang [**L4**], where however an error occurs: known equivalent representations are counted twice). The *Weil representation*, which appears in Section 5.3, plays here a special rôle in that it is a projective representation of $SL_2(\mathbb{R})$.

Induced representations. With the use of induction, one may construct, in great generality, representations of the group G from representations of its subgroups. To this end, let $H \subset G$ be a closed subgroup and $K \subset G$ a subgroup such that

$$H \times K \to G = H \cdot K$$

is a topological isomorphism. Further, let

$$\sigma : H \to \operatorname{Aut} V$$

be a finite–dimensional continuous representation. Then the associated *induced representation*

$$\tau = \mathrm{Ind}_H^G \sigma$$

of G is given, in that G operates by right translation on a space $\mathcal{H}(\sigma)$, which is spanned by the functions

$$\phi : G \to V$$

with

$$\phi(hg) = \Delta(h)^{1/2} \sigma(h) \phi(g) \quad \text{for all } h \in H,\ g \in G$$

and

$$\| \phi \|^2 := \int_K |\phi(k)|^2 dk < \infty,$$

where $\Delta = \Delta_H$ is the *modular function* of H, that is, the function with

$$d_r(xy) = \Delta_H(x)\, d_r y \quad \text{for all } x, y \in H,$$

whenever $d_r y$ is a right–invariant Haar measure on H. The factor $\Delta^{1/2}$ works so that τ turns out to be unitary whenever σ is.

One of the most significant theorems from representation theory is the *subrepresentation theorem*, which is a statement that all *interesting* representations are captured as subrepresentations of known induced representations. As an example, the Schrödinger representation is relatively easy to realize as an induced representation (see, for example, BERNDT [**Be**]). The induced representations were particularly strongly propagated by the efforts of MACKEY, [**M**].

The adjoint and coadjoint representations. To every Lie group G there belongs a Lie algebra $\mathfrak{g} = \mathrm{Lie}\, G$, and also a natural representation of G on $\mathfrak{g}$, called the *adjoint representation* Ad. It arises from the fact that through conjugation

$$\kappa_g(g_0) = \varrho_{g^{-1}} \lambda_g(g_0) = g g_0 g^{-1}$$

a diffeomorphism of G is defined, and so, for every $g \in G$, κ_g induces maps of the tangent spaces $T_{g_0}G$. In particular, for $g_0 = e$, T_eG is mapped to itself. It is written as

$$\mathrm{Ad}_g := (\kappa_g)_{*e} = (\varrho_{g^{-1}} \lambda_g)_{*e}.$$

The action

$$\begin{aligned} G \times T_eG &\to T_eG, \\ (g, Y) &\mapsto \mathrm{Ad}_g Y, \end{aligned}$$

is defined via the identification of T_eG and $\mathfrak{g}$, because $\kappa_{g_1 g_2} = \kappa_{g_1} \kappa_{g_2}$ is a representation of G on $\mathfrak{g}$, in which the operators are also written as $\mathrm{Ad}(g)$.

One may also consider the contragredient representation, as in Section D.1, on the dual space $\mathfrak{g}^*$ of $\mathfrak{g}$ to this representation. This is called the *coadjoint representation* of G, and is of particular significance for symplectic geometry.

Since for all $g_0 \in G$, the $\kappa_g(g_0)$ are differentiably dependent on g, the infinitesimal representation $d(\mathrm{Ad}) =: \mathrm{ad}$ associated to Ad can be built as was described above under the infinitesmal method. It is not hard to see (see KIRILLOV ([**Ki**], p. 97)) that then

$$\mathrm{ad}\, X(Y) = [X, Y].$$

Bibliography

[AM] Abraham, R., and Marsden, J.E., *Foundations of Mechanics*, Benjamin–Cummings, Reading, MA, 1978.

[Ae] Aebischer, B., Borer, M., Kälin, M., Leuenberger, Ch., and Reimann, H.M., *Symplectic Geometry*, Prog. Math., vol. 124, Birkhäuser, Basel, 1994.

[A] Arnold, V.I., *Mathematical Methods of Classical Mechanics*, Springer, New York, 1978.

[Ar] Artin, E., *Geometric Algebra*, Interscience Publ., New York, 1957.

[Be] Berndt, R., *Darstellungen der Heisenberggruppe und Thetafunktionen*, Hamburger Beiträge zur Mathematik, Heft 3, 1988.

[BeS] Berndt, R., and Schmidt, R., *Elements of the Representation Theory of the Jacobi Group*, Prog. Math., vol. 163, Birkhäuser, Basel, 1998.

[BS] Berndt, R., and Slodowy, P., *Seminar über Darstellungen von $SL_2(F)$ und $GL_2(F)$, F ein lokaler Körper*, Hamburger Beiträge zur Mathematik, Heft 20, 1992.

[Bl] Blair, D., *Contact Manifolds in Riemannian Geometry*, Lecture Notes in Math., vol. 509, Springer, Berlin, 1976.

[Bo] Bourbaki, N., *Eléments de Mathématique, Algèbre, Chapitre III: Algèbre multilinéaire.* Hermann, Paris, 1948.

[C] Cartan, H., *Differential forms*, Houghton–Mifflin, Boston, 1970.

[Ca] Cartier, P., *Quantum Mechanical Commutation Relations and Theta Functions*, in *Algebraic Groups and Discontinuous Groups* (A. Borel and G.D. Mostow, editors) Proc. Symp. Pure Math., vol. 9, Amer. Math. Soc., Providence, RI, 1966, pp.361–383.

[Ch] Chern, S.S., *Complex Manifolds without Potential Theory*, Van Nostrand, Princeton, NJ, 1967.

[Ce] Chevalley, C., *Théorie des groupes de Lie*, Hermann, Paris, 1951.

[CH] Courant, R. and Hilbert, D., *Methods of Mathematical Physics, Volume I*, Interscience, New York, 1953.

[D] Deligne, P., *Equations Différentielles à Points Singuliers Réguliers*, Lecture Notes in Math., vol. 163, Springer, Berlin, 1970.

[dC] do Carmo, M., *Riemannian Geometry*, Birkhäuser, Boston, 1992.

[E] Eichler, M., *Introduction to the theory of algebraic numbers and functions*, Academic Press, New York, 1966.

[EG] El Gradechi, M.A., *Théories classiques et quantiques sur l'espace-temps anti-de Sitter et leurs limites à courbure null*, Thèse de Doctorat, Université Paris 7, Paris, 1991.

[FL] Fischer, W., Lieb, I., *Funktionentheorie*, 7. Auflage. Vieweg, Braunschweig–Wiesbaden, 1994.

[F] Forster, O., *Analysis 1*, 3. Auflage. Vieweg, Braunschweig–Wiesbaden, 1984.

[FRF] Forster, O., *Lectures on Riemann surfaces*, Springer, New York, 1981.

[Go] Godement, R., *Topologie algébrique et théorie des faisceaux*, Hermann, Paris, 1958.

[GW] Goodman, R., and Wallach, N.R., *Representations and invariants of the classical groups*, Cambridge University Press, 1998.

[Gb] Greub, W., *Multilinear Algebra*, Springer, New York 1997.

[Gr] Gromov, M., *Pseudoholomorphic Curves in Symplectic Manifolds*, Invent. Math. **82** (1985), 307–347.

[GS] Guillemin, V., and Sternberg, S., *Symplectic Techniques in Physics*, Cambridge University Press, 1984.

[HR] Holmann, H., and Rummler, H., *Alternierende Differentialformen*, Bibbliogr. Inst., Mannheim, 1972.

[HZ] Hofer, H., and Zehnder, E., *Symplectic Invariants and Hamiltonian Dynamics*, Birkhäuser, Basel, 1994.

[Ja] Jacobson, N., *Basic Algebra*, Freeman, San Francisco, 1974.

[K] Kähler, E., *Über eine bemerkenswerte Hermitesche Metrik*, Abh. Math. Sem. Univ. Hamburg **9** (1933) 173–186.

[K1] Kähler, E., *Einführung in die Theorie der Systeme von Differentialgleichungen*, Teubner, Leipzig, 1934.

[K2] Kähler, E., *Der innere Differentialkalkül*, Rendiconti die Matematica **21** (1962) 425–523.

[Ki] Kirillov, A.A., *Elements of the Theory of Representations*, Springer, Berlin, 1976.

[Kn] Knapp, A.W., *Representation Theory of Semisimple Groups*, Princeton University Press, 1986.

[L1] Lang, S., *Algebra*, Addison–Wesley, Menlo Park, NJ, 1984.

[L2] Lang, S. *Fundamentals of Differential Geometry*, Springer, New York, 1999.

[L3] Lang, S., *Linear Algebra*, Springer, New York, 1985.

[L4] Lang, S., $SL_2(\mathbb{R})$, Springer, New York, 1989.

[L5] Lang, S., *Undergraduate Analysis*, Springer, New York, 1983.

[LL] Landau, L.D., and Lifschitz, E.M., *The Classical Theory of Fields*, Addison–Wesley, Reading, MA., 1961.

[LV] Lion, G., and Vergne, M., *The Weil Representation, Maslov Index and Theta Series*, Prog. Math., vol.6, Birkhäuser, Basel, 1980.

[M] Mackey, G.W., *Unitary Group Representations in Physics, Probability, and Number Theory*, Benjamin/Cummings, Reading, MA, 1978.

[Ma] Marcus, M., *Finite Dimensional Multilinear Algebra (in two parts)*, Marcel Dekker, New York, 1977.

[ML] Mac Lane, S., *Homology*, Springer, Berlin, 1963.

[MS] McDuff, D., and Salamon, D.A., *Introduction to Symplectic Topology*, Oxford University Press, 1995.

[Mu] Mumford, D., *Algebraic Geometry*, I. Springer, New York 1976.

[Mu1] Mumford, D., *Tata Lectures on Theta* II, Prog. Math., vol. 28, Birkhäuser, Boston, 1984.

[Mu2] Mumford, D., *Tata Lectures on Theta* III, Prog. Math., vol. 97, Birkhäuser, Boston, 1991.

[Sa] Satake, I., *Algebraic Structures of Symmetric Domains*, Iwanami Shoten, Tokyo, and Princeton University Press, Princeton, 1980.

[Sch] Schottenloher, M., *Geometrie und Symmetrie in der Physik*, Vieweg, Braunschweig/Wiesbaden, 1995.

[Si1] Siegel, C.L., *Symplectic Geometry*, Academic Press, New York, 1964.

[Si2] Siegel, C.L., *Topics in Complex Function Theory*, Vol. III. Wiley–Interscience, New York, 1973.

[SM] Siegel, C.L., and Moser, J.K., *Lectures on Celestial Mechanics*, Springer, Berlin, 1971.

[So] Souriau, J.–M., *Structure of Dynamical Systems, A Symplectic View of Physics*, Prog. Math., vol. 149, Birkhäuser, Boston, 1997.

[St] Sternberg, S., *Lectures on Differential Geometry*, Prentice–Hall, Englewood Cliffs, NJ, 1964; 2nd ed., Chelsea, New York, 1983.

[V] Vaisman, I., *Symplectic Geometry and Secondary Characteristic Classes*, Prog. Math., vol. 72, Birkhäuser, Boston, 1987.

[Wa] Wallach, N.R., *Symplectic Geometry and Fourier Analysis*, Math. Sci. Press, Brookline, MA, 1977.

[We] Weil, A., *Variétés Kählériennes*, Hermann Paris, 1957

[We1] Weil, A., *Sur certains groupes d'opérateurs unitaires*, Acta Math. **111** (1964) 143–211.

[W] Weyl, H. *The Classical Groups*. Princeton University Press, 1946.

[Wi] Wigner, E., *Group Theory and its Application to the Quantum Mechanics of Atomic Spectra*, Academic Press, New York, 1959.

[Wo] Woodhouse, N., *Geometric Quantization*, Clarendon Press, Oxford, 1980.

Index

Symbols

Symplectic vector spaces

$(V,\, \omega)$	symplectic vector space	14
$W^{\perp}$ to $W \subset V$	the orthogonal space relative to ω	22
$\operatorname{rad} W = W \cap W^{\perp}$	radical of W	22
$W^{\mathrm{red}} = W/\operatorname{rad} W$	the symplectic space associated to W	23
$L = L^{\perp} \subset V$	Lagrangian subspace	23
$\mathcal{L}(V)$	collection of Lagrangian spaces $L \subset V$	26
$\mathcal{T}(L)$	$= \{L' \in \mathcal{L}(V),\, L \oplus L' = V\}$	27
$\mathcal{J} = \mathcal{J}(V,\, \omega)$	space of ω–compatible positive complex structures J	36
$\mathfrak{H}_n = Sp_n(\mathbb{R})/U(n)$	Siegel upper half–space $\simeq \mathcal{J}(\mathbb{R}^n,\, \omega_o)$	36
$h(v,\, w)$	$= g(v,\, w) + i\omega(v,\, w)$ hermitian, Riemannian and outer form	32, 55
$J \in \operatorname{Aut} V$	complex structure ($J^2 = -id$)	29

Symplectic manifolds

$m \in M$	point of a differentiable real manifold of dimension $2n$	
$\varphi : U \to \mathbb{R}^{2n}$	chart of a neighborhood $U \subset M$	144
$\varphi(m) = (q,\, p)$	symplectic standard coordinates	40

$f \in \mathcal{F}(M)$	differentiable function on M	147
$X \in V(M)$	differentiable vector field on M $(= \Gamma(TM))$	155
$\alpha \in \Omega^q(M)$	differentiable exterior q–form on M	161
$\omega \in \Omega^2(M)$	symplectic form on M, in particular	39
$\omega_0 = \sum dq_i \wedge dp_i$	the standard form	40
$\vartheta = \sum p_i \, dq_i$	the Liouville form	49
$\omega^\#$	fundamental duality $\Omega^1(M) \xrightarrow{\sim} V(M)$	78
ω^b	inverse mapping to $\omega^\#$	78
$F : M \to M'$	diffeomorphism $m \mapsto F(m) = m'$	146
F_{*m}	mapping of the tangent vectors $F_{*m}(X_m) = X'_{m'}$	151
$F^*_{m'}$	mapping of the 1–forms $F^*_m(\alpha'_{m'}) = \alpha_m$	151
$i(X)\alpha$	inner product of $X \in V(M)$ with $\alpha \in \Omega^q(M)$	17, 78
$L_X f$	Lie derivative of $f \in \mathcal{F}(M)$ by $X \in V(M)$	77
$L_X \alpha$	Lie derivative of $\alpha \in \Omega^q(M)$	
	by $X \in V(M)$	56, 164
$L_X Y = [X, Y]$	Lie derivative of $Y \in V(M)$ by $X \in V(M)$	78, 173
∇	connection	166
$\nabla_X \alpha$, $\nabla_X Y$	covariant derivative	166
F_t	flow of $X \in V(M)$	45, 165
$\{f, h\}$	Poisson bracket of $f, h \in \mathcal{F}(M)$	85, 86
$X_f \in \mathrm{Ham}\,(M)$	Hamiltonian vector field to $f \in \mathcal{F}(M)$	80
X_M	infinitesimal generator of $X \in \mathfrak{g} = \mathrm{Lie}\, G$	
	and the group operation ϕ	99, 103
$\phi : G \times M \to M$	with $\phi(g, m) = gm = \phi_g(m) = \psi_m(g)$	99
$\rho_{g_0} = gg_0$	right translation of g with g_0	174
$\lambda_{g_0} = g_0 g$	left translation of g with g_0	174
$\kappa_{g_0} = g_0 g g_0^{-1}$	conjugation with g_0	57
Φ	moment map	100

Representations

G	Lie group	173
$\mathfrak{g} = \mathrm{Lie}\, G$	associated Lie algebra	174
π	continuous representation of a Lie group G	187
π^*	associated contragredient representation	186